Alec Crawford has spent his whole career in the world of marine salvage in various locations around the world, including the UK, USA, Vietnam, South Africa and all through the Mediterranean. Actively involved in salvage technology, he developed an environmental oil removal tool and a novel winder system which was highlighted in the Timewatch documentary *The Lost Liner and the Empire's Gold*. His company, Deep Water Recovery and Exploration, was one of the world's leading salvage businesses.

TREASURE ISLANDS

True Tales of a Shipwreck Hunter

Alec Crawford

BIRLINN

First published in Great Britain in 2020 by
Birlinn Ltd
West Newington House
10 Newington Road
Edinburgh
EH9 1QS

www.birlinn.co.uk

ISBN: 978 1 78027 601 4

British Library Cataloguing-in-Publication Data
A catalogue record for this book is available on request from the British Library

Papers used by Birlinn are from well-managed forests
and other responsible sources

Typeset by Initial Typesetting Services, Edinburgh
Printed and bound by Clays Ltd, Elcograf S.p.A.

Contents

List of Illustrations

Dan, Isaac, John, Peter and Alec on the *Vesper*.
The *Dewy Rose*, Northbay, Barra.
Peter preparing sacks of scallops on Castlebay pier.
Alec and Chris studying the *Adelaar*'s cannon.
Aboard the *Dewy Rose*.
Our morning's haul of scallops, Barra.
Chris with a whisky bottle from the SS *Politician*, Eriskay.
Our Halflinger travels to Shetland.
The mailboat *Good Shepherd 2*.
The team: Alec, Simon and John-Andrew.
Elizabeth, on Foula.
The *Oceanic* at Belfast.
A bearing on the *Oceanic* crankshaft.
Working in among the *Oceanic* engines.
The *Good Shepherd 3* being loaded at Foula.
The *Oceanic*'s anchor used as a mooring for *Trygg*.
Trygg at the entrance to Ham Voe.
The remains of *Trygg* after the storm.
Valorous anchored over the *Canadia*, Fair Isle.
Moya relaxes on *Valorous*.
A deck removal grab designed by Alec and Moya Crawford.

Acknowledgements

My main thanks are given to the Foula folk who accepted us as their own, and to those I have been unable to include who also helped us to pursue our unusual way of life. I hope they enjoyed it as much as we did.

In writing the book I owe most to Josie Jules Andrews, who taught me during my four years of Life Writing classes at Dundee University, where I learned to both enjoy writing as well as the technique of writing a story. Sukey Roxburgh gave me purpose to the book, both encouraging and assisting me to complete the first draft. Help in improving the manuscript along the way was given by Gale Winskill, with a strong critique which I definitely required, and later Sheila Anderson and Bet McCallum for improving my English at different stages of the project.

Hugh Andrew of Birlinn kindly guided me towards the areas of most interest, and thanks to Andrew Simmons and Deborah Warner for the editing.

I would like to thank Merseyside County Museums, Harland and Wolff, Belfast, and the Shetland Museum for all their help over the years in obtaining information and photographs.

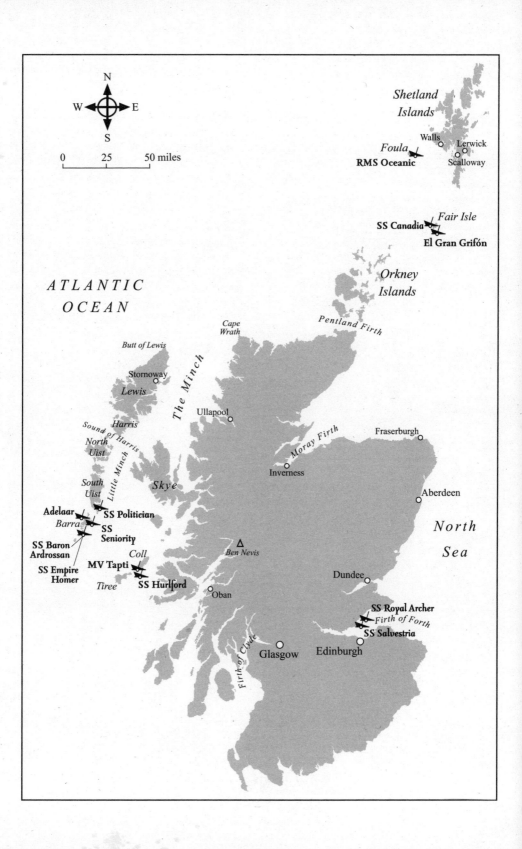

N
W E
S

0   25   50 miles

*Shetland
Islands*

Walls ○   ○ Lerwick
Foula ○
RMS Oceanic   ○ Scalloway

SS Canadia ✈ *Fair Isle*
El Gran Grifón

*Orkney
Islands*

*ATLANTIC
OCEAN*

*Cape
Wrath*

Pentland Firth

Butt of Lewis

Stornoway ○
Lewis   *The Minch*

Ullapool ○

Harris
Sound of Harris   *Moray Firth*   Fraserburgh ○
*North
Uist*
Little Minch   Inverness ○

*South
Uist*   *Skye*

Aberdeen ○

Adelaar ✈ **SS Politician**
Barra
**SS Baron
Ardrossan**   **SS
Seniority**   *North
Sea*

**SS Empire
Homer**   *Coll*   △
Ben Nevis
MV Tapti ✈
Tiree   **SS Hurlford**   Dundee ○

Oban ○

SS Royal Archer
Firth of Forth
SS Salvestria

Firth of Clyde
Glasgow ○   Edinburgh ○

1

Learning from the Wreck
of the *Salvestria*

I set off on a beautiful July morning in 1971 to drive the twenty-five miles from my cottage to Methil on the Firth of Forth. Dan, a local scrap merchant and businessman, had been telling us about a diver named Peter for weeks: how good he was, how he could turn the salvage project around. Since Dan provided the money, I listened. Today we'd finally get to meet him.

As I arrived at the docks, I spotted Dan's neatly painted salvage boat as it lay at the entrance. Her ropes were lying slack, as the onshore wind held her against the pier. The slow, rhythmic beat of the engine was audible in the otherwise silent and deserted dock.

Working on Dan's project was like an apprenticeship for me – and Dan as well. I was paid a small basic wage that was barely sufficient to live on and a percentage of the value of the recoveries. And now Peter would be working on the same basis, along with John, the other diver. The project was kept alive by our dreams and Dan's money.

Dan's uncle John ran a diverse business in Peterhead that included scrap, property, fishing-boat equipment, ship-breaking and more. Dan's fervour regarding Peter was based solely on his uncle's recommendation. Plus, Peter had ten years' experience, whereas I had less than one. I had been working on this project for five weeks, only hired by Dan as the diver-in-charge when no one else was available. Secretly, I was relieved to welcome Peter.

When he arrived, Dan was quickly out of his car, brusque and upright in his movements; he had a large grin on his face and was anxious to introduce Peter, who appeared to be ten years older than me. He had the confident look of a diver with a military stature. Dan was right. I felt intuitively that we could work together.

'Alec, this is Peter.'

'Pleased to meet you,' I said, nodding at him, a little too formally probably.

Peter smiled, then said, 'And this is Yak, my dog.'

A dog? I thought. I couldn't imagine her on the boat among all the deck machinery. 'Where'd you get her?' I asked.

'She's an isle of Barra sheepdog,' was the prompt reply.

'Has she been on a boat?'

'Yeah,' said Peter with a flick of his hand. 'She was on the last salvage ship with me.'

I squatted down to stroke Yak. Dan had previously made no mention of her – perhaps he didn't know – but she didn't look as though she'd be a problem.

'What d' you want to do, Peter?' Dan cut in. 'Should we put your gear on the boat? We'll not start till Monday.'

'Anyone staying on the boat?'

'Just Bob, one of the crew. There's a spare bunk and plenty food.'

The day was beautiful, but perhaps the thought of spending the weekend in this derelict dock didn't appeal to him.

'Where d' you stay, Alec?' he asked.

'About twenty-five miles north of here,' I replied, failing to realise the implication of the question.

'Have you a spare room?' he asked.

I had no real plans for the weekend but was not sure my cottage would be up to his expectations. 'You're welcome to stay,' I said, 'but I came on the motorbike. Unless Dan can run you over . . .'

'That's okay,' said Peter. 'I'll jump on the back of the bike.'

'What about the dog?' I asked, taken aback by the haste of the decision.

'I'll hold the dog,' said Peter. 'She'll be fine.'

The shepherd's cottage I was living in was traditional, with three rooms, built from local whinstone. The front door lay in the centre and the pan-tiled roof had been extended in the 1930s to include a kitchen and bathroom. It lay next to a minor road, the traffic partly screened by trees. There were no other houses visible, just some low-lying farm buildings and the remains of two rows of derelict cottages on the higher ground behind. It stood out like a proud survivor, surrounded by willow herb, nettles and uncut grass. As we entered the cottage Peter gave a good-natured apology for asking to stay, saying that he would prefer not to live on Dan's boat but rent his own place in the future.

At the age of twenty-three, I was lucky to have this as a base. I was happy staying there; it gave me freedom, a bolthole to lick my wounds when life was not going well. After paying to train as a diver, a shortage of cash had been a problem, forcing me to make a living with various short-term projects, mostly involving buying and selling, until I found the job with Dan.

I relaxed in a chair and was soon listening to Peter's stories about the places he had worked, my ears pricking up when he talked about the islands. He had spent some years on the Isles of Scilly and then moved to the Outer Hebrides, working on the isle of Barra for a year. The idea of living on a remote island had always appealed to me. I had seen Peter's name in a diving book and had looked it up again after Dan mentioned he would be joining us. It related to his work on HMS *Association*, a treasure wreck lost in 1707 on the Isles of Scilly.

'Interesting work?' I queried, mentioning the *Association*, assuming he had made a lot of money on it.

'Yes, but we made more money crayfish diving. The wrecks were a lucky find – they were in the area where we were searching for crayfish.'

'Why'd you leave?' I asked, thinking of this diving paradise.

'Ran out of shallow water crayfish. You'd have loved the diving. Clear water, not like the Firth of Forth.'

'It's something I dream about, working undiscovered wrecks in crystal-clear water.'

'They're there on Barra and I'll be going back,' replied Peter, with a positive flick of his head and a smile.

Going to the local pub in the evening, I was surprised by Peter – perhaps it was realising my youth and inexperience, or jealousy maybe? But I watched him woo the people and turn their heads. Whether he was a nightmare or a godsend, I didn't know. I sat in silence, intrigued by this stranger. *Dan was right*, I thought. *If Peter's as good at his job as he is socially, the man's a winner.*

Returning to the boat after the weekend, we made our way down the ladder, Peter carrying Yak against his chest.

The *Vesper* was a beautiful boat, with her neatly painted hull and attractive lines. She was a typical old fishing boat, built in 1911 with sails and later fitted with an engine. At 45 feet long, with a small wheelhouse aft, she was small for salvage. Her main attraction for Dan would have been the price: the fishing industry was doing well and good working boats were fetching high prices. The *Vesper* would have been a bargain. The navigational equipment was sparse, with only a magnetic compass and an old echo-sounder; for shore communications she had an old Sailor VHF radio. Whether anyone had a licence to use it I wasn't sure, but it worked perfectly well without one.

The Firth of Forth, in which Methil lay, was said to contain hundreds of shipwrecks, from the small *Blessing of Burntisland*, lost in 1633 (said to be carrying Charles I's valuable personal possessions) to the 12,950-ton *Campania*, originally a Cunard passenger liner built in 1893 and a holder of the Blue Riband, an accolade given to a passenger liner crossing the Atlantic in regular service with the record highest speed. She had been purchased by the Royal Navy in 1914 and converted to an aircraft carrier but had dragged her anchor during a gale in 1918, causing her to collide with the battleship HMS *Royal Oak*, which pierced her hull, then striking HMS *Glorious* before she sank.

Dan had bought some wrecks from the original insurers: the *Salvestria*, at nearly 12,000 tons, was the largest of them. This had been a financial punt based on the vast quantity of copper and brass lying on these old wrecks in the form of the engine-room components and other fittings that had substantial value. 'The scrap's worth several hundred thousand,' said Dan, before qualifying it with, 'If we can get it up.' This was at a time when you could buy a house for a few thousand pounds.

Salvage on ships has gone on for centuries; along with gold prospecting, it is one of the few opportunities where an individual, or small group of individuals, can start with practically no capital and have the prospect of making a fortune. The most valuable wrecks contain cargoes of silver or gold, followed by those with metals such as tin, copper and lead. Wrecks of all types have been worked in the past to recover materials of value or to clear them from navigation channels. After the war it was relatively easy money, as the salvors had the pick of the wrecks but they had to work with Standard Gear, the old-fashioned diving suits with copper helmets that were both labour-consuming and restrictive to use. The advent of Scuba gear, with its underwater freedom, opened up opportunities for people like me to search for and recover brass from wrecks with people like Dan who were prepared to risk their capital in backing us.

Peter looked up from Dan's chart. He had already seen most of the information at the cottage, where I had given him a brief explanation of our work.

'What do you think, Peter?' asked Dan.

From the fuss Dan had made over Peter coming to work, his answer would carry a lot of weight. I watched his right hand twitch . . . I thought it might be the result of a diving accident. He looked at Dan, paused as if to build tension, and then replied, 'No problem, Dan. It looks an easy job.'

Dan's face lit up. He turned to face me. 'I told you, Alec, we'll all make our fortunes.'

I grinned, delighted to be part of a winning team, but a few doubts remained.

John and I had initially dived on the *Salvestria* before Peter arrived, but the visibility was almost zero, caused by silty water from the winter storms coming down the estuary. A local trawler skipper told us we should wait for a month or two and it would improve. Taking his advice, we moved to other wrecks. Peter would join us to salvage these first, then we would return to the *Salvestria*.

On Peter's first trip we sailed out to the wreck of the *Royal Archer*. I was buoyed up with hope, as I always enjoyed the trip out: the compressor filling the air bottles; diving suits brought on deck to be checked; the general bustle as the boat's derricks and winches were prepared for lifting. No orders were given: everyone knew the part they had to play. I watched cautiously as Peter's air bottles charged; they were ex-aircraft oxygen cylinders that had their wire bindings removed, reducing their weight and also their strength. They were still common in the diving world, often referred to as 'bombs'. *With good reason*, I thought.

Peter made light of it, but I fetched an old rubbish bin, filled it with water and sat the bottles in it while they filled. This would limit the damage if they exploded. Yak sat watching the compressor, waiting for me to blow moisture from the system, when she would try and catch the water in her mouth.

'What d' you think, Yak? Are they safe?' I asked, before checking myself and smiling.

When all the equipment was ready, Bob brought tea on deck and Peter regaled us with stories about his work on the isle of Barra, adding that he looked forward to returning when we had finished this job.

'It was a good job at the shellfish factory,' he said. 'I managed part of the company, as well as looking after the lobster cages. The potential was enormous, and in my free time I dived for scallops, which the factory bought from me.'

'Why did you leave?' John asked.

'The owner was killed in a light aircraft crash flying from Barra to the mainland.'

'I'm sorry,' I said, surprised this should have affected the factory, if it had so much potential. 'What happened to the factory?'

'It closed. They're still sending lobsters out, but only in a small way. If I go back it'll be with my own boat, to work on salvage and dive for scallops.'

'Are there many wrecks?'

'The island's littered with them. Most have never been worked. With the right equipment I could make a fortune.' Peter rubbed his hands together. Yak, who had been sitting quietly staring out to sea, looked up to check if he was signalling to her. He bent down to pat her.

Peter had sown the seeds in my mind and we were getting on well together, making me wonder if I might be in with a chance of going to Barra with him. It would be too good an opportunity to miss, as I had no other work after this summer with Dan, we just had to make enough money from this project to buy a boat – that is, if Peter would take me. I was brought back to reality when

Dan leant out of a wheelhouse window and asked, 'Peter, would you like to see the wreck?'

Peter squeezed into the wheelhouse; the sounder was on the left of the ship's wheel. It made a clicking sound each time it marked the depth on a strip of paper which fed from one roller to another. Dan had been given the position of the wreck by fishermen who had caught their nets on it unintentionally. They had given him a sketch of buildings and prominent landmarks on the shore, which he lined up until the echo-sounder confirmed the wreck by displaying an abrupt change in depth.

'There she is,' said Dan, cutting the throttle to stop the boat as a small grapple was thrown overboard by Bob, the rope snaking over the gunwale until the grapple landed on the seabed, or hopefully the wreck. We watched as the tide took the loose coils of rope, the buoy spinning wildly before it was pulled firmly by the current as the rope became taught. Below the buoy was the wreck of the *Royal Archer*, a general cargo ship of 2,266 tons built in 1928, sunk after striking a mine in 1940 while carrying cargo from London to Leith.

It is easy to imagine great big ships resting on the seabed just as they would have looked floating on the surface, but in shallow water, where there is no time, money or equipment available to lift them, ships are often blown down with explosives to keep them clear of traffic passing above them. Sometimes this work is carried out by the navy shortly after the loss of the ship, or later by salvors, who also take the opportunity to recover anything of value. The result is a mass of twisted steel on the seabed, with few distinct shapes to indicate the original structure, and over time, as the ship settles into the seabed, access becomes blocked by mud and silt accumulating within it.

John and I had been down to see the Royal Archer on previous visits and knew the stern remained reasonably intact, acting as our starting point to identify other parts of the wreck. As I swam down beside the mooring rope, the suspended sediment increased and the light became less as I ventured deeper until I had to go hand over hand down the rope to keep sight of it. Landing on the stern of the wreck, I would stay there for a few minutes to relax and let my eyes acclimatise to the darkness, my vision generally restricted to 6 feet. It was worse than working on a dark night, giving me a lonely, eerie but exciting feeling, heightened by the hiss of air as I sucked in – as if I were taking a breath through gritted teeth.

Swimming along the rail at the starboard side of the wreck, the current had scoured the sandy seabed alongside the hull, leaving a trench five feet deep. The occasional conger eel made a dash for cover when disturbed and six-inch-long butterfish that looked like small eels played among the wreckage. There were patches of anemones and dead man's fingers, a soft coral that was aptly named due to its shape and texture; it felt like touching flesh if your fingers brushed it unknowingly. The whole scene was dull grey; my torch was of little use in the dirty water as the light was reflected back.

After the ship's aft hold, the wreck lost its shape and became a jungle of twisted steel plates. This is where I estimated the engine room lay, the area to search for copper and brass. Every time I tried to move something the silt clouded up like smoke, reducing the visibility around me to zero. I often felt my breathing rate increase and my heart beat harder, the adrenalin pumping through my body as I became alert to the possibility of being trapped under wreckage. It paid to be careful: a wreck provided a rich home for fish, making it tempting for fishing boats to trawl as close as possible. The snagged nets floated up from the wreck like a curtain, and in shallow water they moved gently backwards and forwards in the tide, catching anything that drifted their way, including divers.

Unrecognisable parts of the ship would tower above me; it's

like passing a tall building in the dark. Initially, when I started this work, if a large structure loomed over me it would make my heart stop. If divers exist who have no fear, I am not one of them.

My job was to look for the larger pieces of brass, such as condensers weighing several tons, pumps and bearings, but if I failed to find them, any brass valve or copper pipe was quickly attached to a wire rope and winched to the surface. As brass is pulled off a wreck, steel plates move, causing the visibility in the area to become impenetrably black. A diver has to keep well clear. Sometimes a loose piece of plate would catch on the wire and fall when it was halfway up to the boat. We preferred swimming to the surface even when we had to return below, particularly if the object was large or looked to be stuck fast in the wreckage When the load came on it might suddenly spring out, making the visibility zero and giving the diver no chance of getting out of the way or seeing it fall as it was being hauled up.

That day, the *Vesper* was anchored up-tide of the wreck, as Peter prepared to enter the water. I remained on deck but was keen to dive with him, though he preferred to have a look on his own. Clad in his wetsuit with a bottle on his back, Peter entered the water with a neat forward somersault. We were all impressed. We settled on deck with a second mug of tea in our hands, waiting for his return. I thought how different the diving would be on the isle of Barra because of the beautifully clear water. I longed to go there. We had sat in the pub over the weekend discussing wrecks on the island, the beautiful sandy beaches and the cottage he knew that a crofter would rent to him. It was set in among inlets and small islands, providing an ideal place to moor a salvage boat. He talked of the nice folk that inhabited the island, the dances, social life and money that could be made off the wrecks. I was hooked.

Dressed in my suit as the standby diver, I watched his bubbles rise to the surface. After five minutes Dan checked his watch, ticking off the minutes, hoping he would not exceed the decompression time. Dan was a worrier. Peter had a single air bottle, so it was unlikely he would have more than twenty minutes at the bottom depth of 95 feet before he ran out of air. John and I had pressure gauges on our bottles to give us an idea of the air remaining, but not Peter. 'I've no problem knowing when my air's running out,' he'd told us.

If we exceeded our bottom times we had to hang on a rope beneath the *Vesper* to decompress. Decompression is like opening a bottle of champagne: opened slowly, no bubbles come to the surface, but open it quickly and the bubbles foam up, blowing the cork out. Human cells function in a similar way: underwater the body is under high pressure and the gases from the diver's breathing are absorbed into the diver's tissues. By surfacing too quickly the deep-water pressure reduces so fast that these gases expanded and, as in the champagne example, they come out of solution as minute bubbles, causing damage to the cells – commonly known as 'the bends'. If we kept to the standard decompression tables and always came up within the correct times, the chance of having a 'bend' was minimal. Our greatest risk was from an accident, such as being trapped underwater due to the dangerous condition of the wreck, particularly after blasting.

Yak, distressed at seeing Peter disappear into the water, stood with her rear legs on the deck and her front legs on the low rail that ran round the *Vesper*. She peered intensely at the bubbles, her head following their movement as Peter swam 95 feet below. Her ears would prick up, her mouth open, with an occasional pant, as she waited for her master to emerge. Peter had assured us she would never jump into the water. 'The Barra folk were crofters as well as fishermen and didn't want their dogs jumping overboard,' he explained. 'As a puppy we held her underwater just long enough for her to dislike it!' It had clearly worked with Yak.

'When is he going to come up?' asked Dan, who was becoming

agitated at the time Peter had been down. Glancing at his watch a few minutes later, he said, 'He should be up by now. His air can't last that long.'

'He can't be out of air, his bubbles are still coming up,' I remarked. I was also getting concerned, though I was impressed by the length of time he had been down with such a small bottle.

When he broke the surface not far from the boat, we all rushed to the rail. I threw a rope to pull him alongside. He climbed up the diving ladder and sat down on the deck. After he took his gloves off Bob handed him a cup of tea and a Tunnock's caramel wafer. Yak sat beside him, eyeing the biscuit. His report was similar to my own. Visibility underwater was up to 15 feet in places, in others it was much less, down to five feet. He could see the stern was reasonably intact but was sunk into the muddy seabed.

'What d' you think, Peter?' asked Dan.

'It's a difficult one. The propeller's below the mud level. Not possible to see if it's cast iron or bronze.' A bronze propeller would be worth several thousand pounds in scrap and be easy to recover, as explosives would cut it off the shaft.

Peter continued giving Dan an appraisal of the wreck, but I could see Dan was disappointed; he had been hoping for a positive report. Unable to see the wreck, he had to believe what he was told – it was a case of blind faith. Secretly, I was reassured – John and I had told Dan much the same, so I knew now we had not been incompetent.

'I'll have a break and go down again,' said Peter.

'You haven't much time left,' shouted Dan.

Peter relaxed before changing his bottle using the quick-release on his backpack and then slipped quietly over the side. After throwing the marked decompression rope and its weight into the water, I sat on the hatch thinking what I would send up from the wreck. The sun came out, cheering us all up while we watched Peter's bubbles. They were almost hypnotic. Dan was continually looking at his watch, and below in the cabin we could hear Bob clattering among the chipped crockery. The only other sound was the slopping of the bilges as the *Vesper* rolled back and forth. It was a peace only experienced at sea.

The tranquillity was suddenly lost as a surge of bubbles appeared, followed by Peter's air bottle breaking the surface of the water. It rose out of the water slowly, rolling over and splashing as it fell back. It lay there, bobbing gently, emitting a slight hiss from the valve as air leaked out. Realising Peter was not attached to it sent a shudder through me. There was panic on deck.

'D' you think something's fallen on him?' Dan asked.

'Don't know,' I replied. 'He might have caught underneath something and it's pulled his bottle out,' I shouted, heading for the side of the boat. John and I, already dressed in wetsuits, grabbed a set of bottles each, then checked our air supply prior to jumping in.

'Bob, Bob! Get up on deck, quick!' called Dan, as he ran towards the engine-room hatch to start the engine.

More than a minute must have passed since the bottle came up, when I heard a shout. 'Anyone seen my bottle?' The call came from the water. We all froze.

'You okay?' I shouted, before picking out Peter on the surface. He turned round to look for his bottle. Astonished but relieved, I pointed to the bottle bobbing about 20 feet from him. Peter swam across and, using the quick-release handle, secured it to his backpack, performed a very neat dive, just like a seal, then disappeared beneath the waves. We all looked at each other, not quite sure what to do.

'Bloody hell!' I exclaimed.

'Did he come all the way from the bottom without his air bottle?' Dan asked, looking at the surface of the water where Peter had disappeared.

'Yes,' I replied. 'The bottle's so light it must have shot to the surface when it slipped out of his backpack. God knows how he managed to swim up.' I shook my head, not sure if it was luck, skill or sheer bravado that kept Peter going.

'Anyone seen my bottle?' mimicked Dan. We all laughed.

By the following weekend the story had started to grow legs and the bottle was now shooting six feet clear of the water. When Peter looked at me in the bar, I shook my head and smiled. I watched him, often with intrigue and some envy, as he entertained and courted. I was getting to know him and his skilful, entertaining ways. He usually sat on a cushioned seat that ran round the inside of the bay window. Yak would sit on the floor next to him. With a table in front of the seat and the bar at one side, it was a popular corner. I would sit quietly, nursing a pint of beer, often with Mary beside me, who I was going out with, while Peter captivated the farming fraternity. Yak, being a sheepdog, seemed to hint at Peter having a connection to the countryside. Peter, who had told me he had no experience of agriculture, would listen intently and ask some pointed technical questions. He had an ability to pick out the key words in a conversation and remember them, using them again when talking on the same subject to give the impression he was an expert. He was a magnet to some women too because he and Yak were fun to be with.

'Why's she called Yak?' I asked, during one of our evenings in the pub, when things had quietened down and I was the last of his audience.

'She's named after a girl,' he replied.

'It's a girl's name?' I queried.

'It's Kay backwards.'

'Who's Kay?'

'A-ha,' he replied, winking. That was a story for another night.

Peter lived life to the full. Whatever money he had, he spent. We were very different when it came to girls as well as lifestyle. He only showed interest in women who came to him; he never actively sought any. Almost ten years younger, I was undoubtedly far less experienced, and not only in terms of diving.

Back on the boat, the days were good. I was gaining experience and growing in confidence. Each day was different, presenting new hope for success, convincing me that salvage was a game of patience and optimism. It was always 'jam tomorrow', but as work progressed I realised that Peter was no better at salvage than John or me. He was a fantastic diver – there was no weather, poor underwater visibility or currents that would restrict him. The limiting factor was the boat on the surface: whether it could keep position, or was able to lift. Diving provided the method for getting to work, but without technical knowledge there was little real work that could be accomplished. I would never be as competent a diver as Peter, but I had a better grasp of engineering and that was the key to salvage. If I wanted to go to the isle of Barra with Peter to work wrecks, it would be reassuring if I learned a lot more about salvage. To this end I bought every relevant book I could find on salvage and steam engines – the engines with the most brass for recovery.

Meanwhile, the work at sea became routine, and with dogged persistence we started to build up some recoveries. But life was never dull with Peter. He was rarely bashful on the boat and sometimes after a long, cold dive he would take off his wetsuit and run round the deck naked, shouting, 'Nik, nik, nik!' An excited Yak would bark and jump in the air as she followed him. This, he exclaimed, was to warm himself up.

One windy day, sitting in my cottage drinking coffee, we discussed our future plans.

'Would you like to go to Barra?' Peter asked.

I was not surprised by the question. He had dropped hints, seeing how interested I was in his stories of his time on the island, but this was the first time he had asked me directly. During previous conversations on Barra, I had assumed I would be going with him, but the way he asked me the question that day made me think he had made a decision: this was no longer rhetoric, it was going to happen.

'Yes,' I replied. Then, thinking my response lacked enthusiasm, I added, 'That'd be great. It's exactly what I'd like to do.'

'Where d' you think we can get a boat? Something we can afford,' he asked.

Ah, I thought, this is it, the boat.

'I know of one,' I said. I had seen the *Dewy Rose* earlier that summer. 'She's a fishing boat, probably affordable, but we'd have to convert her into a salvage vessel.'

'Let's go and see her,' said Peter.

The problem, common to every project I had imagined – from buying and operating a commercial fishing boat when I was sixteen to taking trailer loads of government disposal aircraft propeller blades down to London to sell to antique shops – was money. I knew Peter was short of cash because I had already lent him some to see him through until the scrap was sold. I had saved, but not enough to cover the cost of a boat, its conversion, income for ourselves and the trip to Barra. There was no point trying to borrow from my parents. I had tried this before to purchase a commercial fishing boat. The fallout, when they refused – and I raised the funding elsewhere – had been so great that my father had taken a long time to forgive me.

Nevertheless, one Saturday morning Peter, Yak and I set off in my old Land Rover to make the 60-mile trip to Stonehaven to see the *Dewy Rose*. As well as a fishing town, Stonehaven enjoyed tourist income, with a golf course, open-air swimming pool and a beach over half a mile long. I saw Peter's eyes light up as we mingled with the holidaymakers. Many of them stopped to pet Yak. She liked the attention. Then they would speak to us. This was Peter's kind of place.

At the harbour lay the *Dewy Rose*, an old herring boat that had been converted into a seine netter. She had been flooded by water coming over the pier and now lay against the inner pier, safe but looking neglected. Built in 1929 by Walter Reekie of St Monans, she was 48 feet long with a Scottish-built Kelvin engine. It was half-tide and she was sitting on the harbour bottom, allowing us a good look at the hull before climbing aboard to inspect her through the open hatch.

'Can you get the engine going?' asked Peter.

'Yes,' I said. 'I asked about it last time. They've overhauled it since she sank.'

'D' you think we can buy her?'

'Let's ask the harbour master.' I was sure this was the boat – the start of my future. How could it go wrong? We just had to work harder to make the money to buy her.

We used the derrick on the *Vesper* for lifting small recoveries over the side. For heavy lifts the two winch wires were redirected to run close together either side of the bow. We could not lift anything aboard, but when it reached the surface we could transfer the load to the derrick. If it was too heavy for the derrick, we could leave the object hanging at the bow.

One calm, warm day I was sitting on the boat's rail enjoying the sunshine while holding the signal rope. Peter had dived to shackle on a brass crane slew ring. The lift was expected to be extremely heavy, requiring Dan to use both winch drums. Dan lowered the two wire ropes over the rollers at the bow; Peter attached them to the slew ring that had been blasted free, 90 feet below. On deck the wire ropes ran from the bow, either side of the forward mast onto the two winch drums. Dan stood between the wires to operate the mechanical levers that controlled the winch. I felt the tug on the signal rope from Peter to tell Dan to haul in the wires. He pushed the two winch levers forward, locking in the drive clutches, making the drums turn. As the weight increased I felt the bow of the boat being pulled deeper into the water.

'It's a big lift, Alec,' shouted Dan. I got up from my seat on the rail, tied the signal line and walked forward to look over the bow.

'The bow's down at least three feet,' I replied. 'It's maybe too heavy.'

'That's the brake on,' said Dan, as he ducked under a wire to come forward for a look. 'It should be okay.'

The boat felt quite odd, the load on the bow causing the stern to rise out of the water. Assuming the lift was still attached to the wreck, we hoped the movement of the boat would loosen it. After five minutes the *Vesper* gave a jolt, indicating the slew ring had been pulled clear. Dan started hauling in on both winch drums.

'They're only just lifting it,' shouted Dan. 'It's the biggest lift we've had.' Excited by the prospect of several tons of brass, he shouted to Bob, who was in the wheelhouse, to speed the engine up, needing more power for the winch.

The winch gearbox growled with the heavy load. The wires creaked with the tension, making cracking sounds as they wound onto the drums. Dan looked warily at the wires passing either side of him. They vibrated with the tension. I moved aft to speak to Peter, who had come to the surface.

'It's a big lift,' I shouted. He gave me the thumbs-up, swimming forward to wait for the brass slew ring to reach the surface. Moving amidships, I leaned over the side in the hope of seeing the lift as it surfaced. This lift would be too heavy to transfer to the derrick; it would have to be secured at the bow, enabling us to steam into Methil, where we would use a crane.

'Stop! It's at the surface,' I shouted to Dan. I could clearly see a large section of steel shaft hanging off the slew ring as it broke the surface. This accounted for the extra weight. Dan hauled hard on the brake levers, but the load started to creep downwards. The brakes proved inadequate. From a slow start, the winch drums gathered speed as the wire pulled out. Smoke poured off the drums.

Peter, seeing the load drop, swam away as fast as he could. I looked at Dan, who was attempting to physically force the winch brakes to hold. He finally gave up. Letting go of the useless brake levers, he looked to see if he could get out from between the thrashing wires but realised it was impossible. He crouched down on the deck behind the winches. I knew if one of the wires caught, it would rip the winches right out of the deck, taking Dan with them.

There was nothing I could do. I ran aft and lay flat on the deck alongside the wheelhouse beside John. As the slew ring raced to the seabed, the winch speed increased well beyond any normal speed. The rollers screamed. Grease flew off the over-tensioned wires, wrung out of them like water squeezed from a sponge. Loud cracking noises filled the air as the wires, near breaking point, cut down into the remaining wire on the drums.

At last the slew ring landed on the seabed. The load came off the wires, but both winch drums continued to spin. With no weight, the remaining length of wire flew off the drums in great loops twenty feet high, ripping off any protruding structures, such as lights, crosstrees and aerials on the forward mast. The sound was horrendous – like a large felled tree hitting the ground and scattering its branches. I raised my head to sneak a look towards Dan. He wasn't moving. Debris fell all around him. I was sure the looping wires must be touching him.

Unsure what to do, I lifted myself off the deck, the noise gradually diminishing as the load went off the wires. There was a loud crack as the ends of the wires were torn out of their securing clamps. The noise then changed to the whirr of the empty drums while they continued to turn under their own inertia. At last they stopped and the only sound was the rumble of the engine.

I rushed forward to see if Dan was alive. Peter shouted from the water for us to help him aboard and John gave him a hand. Dan didn't move. Bob cleared some light debris off him. His jersey was streaked with grease from the wires, his hat pulled over his face.

'Dan!' I shouted, as I touched his shoulder.

He started to move. 'Has it stopped?' he asked.

'Yes,' I replied. 'Are you okay?'

We helped him up. He was as white as a sheet.

'The boat's not holed?'

'Just damage on deck,' I replied.

Bob was like a gentle giant as he helped Dan into a sitting position while I scrutinised him for any obvious injuries. Peter and John joined us forward to see what had happened. We were all shocked, the atmosphere subdued. I was still unsure whether or not Dan had been unconscious. He was completely white, as if all the blood had been drained from his face.

'I'll take her back to Methil,' said Bob, going back to the wheelhouse.

As Bob slipped the mooring, Dan sat with a mug of tea in his hand. There were long silences on the way back. Even Yak felt the tension, frightened to go forward and, instead of sitting next to Peter and staring at him, she curled up on a coil of rope next to the steering gear. The incident pulled us closer. To keep ourselves busy we tidied up the mess on the deck.

Dan started to get some colour back – he had been lucky. When the boat tied up, we were a crew in shock, though relieved to come away unscathed.

By Friday evening it was back to the hotel bar as if nothing had happened, but I knew it was a warning, something we had to learn from and none of us wanted to go back to the *Royal Archer*.

We muddled on, diving on various wrecks, recovering two bronze propellers, one blown to small pieces because of our inexperience, but our lifting techniques improved and became safe. There were no methods of knowing what weight we had on the winch at that time and the operations were becoming more of an art than a science as we learned the *feel* on a winch or derrick.

It was a happy team that returned to the largest of Dan's wrecks, the *Salvestria*, with the fresh hope that a new wreck brings. The air compressor, with its petrol engine, worked away noisily as it charged the air bottles on the afterdeck. Dan was at the wheel, with Bob below frying up bacon for sandwiches. Yak looked down the hatch, attracted by the smell, while Peter, John and I debated which section of the wreck we should search.

The *Salvestria*, at 11,938 GRT, was 500 feet long, built in Belfast as the steamship *Cardiganshire* in 1913. In 1929 she had been sold to Christian Salvesen's South Georgia Co., who converted her into a whale factory ship, renamed her *Salvestria* and fitted her with large-capacity tanks to contain whale oil. This made her an easy conversion into an oil tanker during the Second World War. Inward-bound from Aruba, an island off Venezuela, with 12,000 tons of fuel oil in her cargo tanks, she arrived off Methil on 27 July 1940. Steaming towards Inchkeith Island, a magnetic mine detonated under No. 7 and No. 8 tanks, aft of the engine room. *Salvestria* continued forward for 75 yards before her stern touched the seabed. Her bow stayed afloat for about twenty minutes. Oil from her tanks washed ashore on both sides of the Forth. In today's terms this would have been a major environmental disaster, particularly for the Isle of May, with its large breeding colony of puffins, and the Bass Rock, with the world's largest colony of northern gannets. At the time it was accepted as one of the consequences of war – nothing would have been done, and nature would have been left to disperse the oil. As recently as 1967, when the SS *Torrey Canyon*, one of the first generation of supertankers, ran aground between the Isles of Scilly and Land's End in Cornwall, government ministers decided the best way of stopping the oil pollution was to bomb and sink the stricken vessel. The impact on the environment from the spill is still being felt more than fifty years on.

Today, every salvage company's first priority is the removal of oil. Using specialist environmental oil recovery equipment it is possible to attach a pump remotely onto the outside of a ship's tanks and remove the oil within a very short timescale, regardless of depth.

The *Salvestria* became a major navigational hazard. Over the years, she had been bombed with depth charges and swept with wires to reduce her height, taking account of the increased size of ships that required safe passage over her. Nature had also taken its course, as she gradually sank into the soft seabed, the blown-open hull filling with silt.

When the *Vesper* was anchored up-tide of the wreck, we knew there would be no recognisable ship beneath us. Peter carried out his trademark forward flip and disappeared beneath the surface. We waited, more relaxed this time, since the wreck was relatively shallow at a maximum of 80 feet. Peter had an amazingly low air consumption. I assumed it was the result of him being so relaxed – that may also have contributed to him not getting bent when he exceeded his decompression times. I wondered if the bends was partly related to air consumption: if so, I realised I was vulnerable.

Yak, as usual, was peering intensely at the bubbles. When they drifted out of sight, we realised just how huge this wreck was. Peter finally broke the surface, a good distance up-tide of the boat. We pulled him in, leaning eagerly over the rail to hear anything he might say. He kept us in suspense, saying nothing until settled on deck with the usual cup of tea and caramel wafer. Sadly, his report was not favourable. The visibility was worse than it had been on the *Royal Archer*, and the wreck was so broken up that it was impossible to distinguish the different parts of the ship. He had not seen anything worth lifting, although the wreckage stuck up 35 feet above the seabed. After a short rest, Peter changed his air bottle and dived again. He had left us disappointed, but it was a big ship and required many more dives to explore it.

Finding her steam engines seemed impossible, but after a few more days' work we discovered an area containing numerous copper pipes a foot in diameter and twenty feet in length. We hauled them out of the wreckage with the winch wire, and as each one was extracted more would appear. At last we thought our luck had changed when we lifted a pipe so heavy that we had to lash it alongside just under the surface. It was the full length of the boat. Using the ship's radio, Dan organised a small crane to meet us at the entrance to the docks. We were getting so many copper pipes that Dan had become convinced that we were working in part of the engine room.

'There'll be more to remove,' he said, 'before we get to the valuable steam condensers.'

'The condensers are worth a fortune,' Peter informed us. 'There'll be about ten tons of brass in each of them.'

By the time we reached the entrance to the dock the starboard list caused by the pipe seemed less. Four happy faces with four different estimates of the weight of the pipe watched the crane driver as he looked down from the pier above. The crane took the load. We released the wires from the *Vesper*.

'It doesn't seem that heavy,' shouted the driver.

As the pipe came out of the water we could see hard remnants of silt and mud stuck inside. The weight had dropped from our minimum estimate of three tons to an actual weight of half a ton – the original weight must have been made up of mud and silt, which had been washed out on the trip to the dock. Peter, embarrassed by his estimate, suggested we go to the pub. In the circumstances, it seemed a good idea.

On the next trip out we made a plan to place a large explosive charge as deep as we could within the wreck where we were recovering pipes. Taking the explosives down from the surface, I would pass them to John, who lay under a large plate and who, in turn, passed them on to Peter, who had squeezed down through a small gap to get as deep as he could within the wreckage. On one occasion he even took his air bottle off, holding his breath as he squeezed between two plates to lay the charges. Deep in the wreck, the visibility was absolute zero. John held a rope attached to Peter in order that he could find his way back when he had laid the explosives.

When the charge was set, we drifted back on our moorings to keep the *Vesper* as far from the area as our detonating cable would allow. In 1930, the Italian salvage vessel *Artiglio* was reducing the height of the *Florence*, a wreck containing munitions that was a navigational hazard. They were on their final shot before returning home for Christmas when they paid out the electric cable for firing the shot. During the job they had gradually moved closer to the wreck, as familiarity bred contempt. The original two miles had shrunk down to 300 yards. When the blast went off, it created a hole 200 yards wide. The ship was engulfed, killing twelve of the crew. I was usually overcautious on the distance we lay off, but as I gained more experience and our detonator cable became shorter with use (it would be cut on sharp wreckage), I, like the Italians, reduced the distance from the wreck. The shock wave from the exploding charge thumped at the bottom of the *Vesper* and muddy grey water surged up from the wreck. A few fish came to the surface, which we recovered, but we would have to wait several hours before the visibility cleared.

The blast improved the access to the lower part of the ship, exposing more heavy copper pipes, but there was no sign of the steam condensers we were looking for. Removing heavy decking and steel plating has always been a problem in salvage and it was not until many years later, when we were working remotely in 9,800 feet of water, that we developed a deck crushing tool. Weighing fourteen and a half tons, it tore the plates off and rolled them up like a cigar leaf.

We returned to recover pipes but as we pulled each one clear the visibility reduced to zero, making us continue diving in another part of the ship. It was on one of these occasions that I found a spare propeller blade where the wreckage disappeared into the muddy seabed. It was too heavy for the *Vesper*, so to avoid any risk we asked a fisherman in a stern trawler to lift it for us. The trawler dragged the blade clear of the wreckage before we were able to take it to Methil to be lifted out by a crane. This single blade was worth the same as an old but habitable house on the harbour front of a fishing village, making us all see the potential, but I knew the modest success we were now enjoying was the result of the scrap being available rather than any salvage technique.

As the summer came to an end, Peter's life remained active: he had unlimited energy, but it would have been good if he'd had unlimited funds as well. Like a living Mr Micawber, whatever he earnt, he spent a little more, always sure something else would turn up. He was exceptionally good at enjoying himself, never worrying about the money side or the salvage, or anything to do with it. He just did his best every day we worked. We often discussed the future when we drove to Methil, becoming disappointed when the harbour master from Stonehaven informed us that the owner of the *Dewy Rose* had made no decision. Peter remained determined to go back to Barra after the summer, but we both knew it depended on money.

As the weather started to deteriorate, I could feel the pace of life on board the *Vesper* slowing down: long coffee breaks and a high consumption of caramel wafers were becoming the order of the day. We basked in the heat of the coal stove in the cramped accommodation, our wet and clammy oilskins hanging next to it, where they swung with the roll of the ship, dripping water onto the linoleum. The first to go was Bob, who returned to work on trawlers in the north of Scotland.

'When do you think we'll have to stop the salvage work?' Peter asked Dan.

'I've a chance of a contract for the *Vesper* with some survey people. Are you going to Barra?'

Peter looked at me. We had never decided on a date. We had no boat.

'We'll carry on until your contract starts,' said Peter.

Time was important. Who else had Peter told about the Barra wrecks? I had a nagging worry that someone might beat us to them. The island seemed to act like a magnet to Peter, pulling him towards it – or perhaps it was a girl. It didn't matter which: it sounded like treasure island to me. Considering that Peter had only dived on one of the wrecks on Barra, and of the rest he only knew where they had been sunk, I knew we were both taking a large leap in the dark, but I was convinced it was a chance of a lifetime.

If we were to sail to the Outer Hebrides, the initial expense would rest entirely on my shoulders, but his knowledge, friends on the island and diving skills were essential. We came to an agreement: if I could raise the capital, we would become an equal partnership after half my outlay had been repaid. Nothing was written on paper, we did not even shake hands: we trusted each other. Perhaps it was easy to have so much trust as our lives depended on each other every time we dived.

Peter had no interest in raising money; his diving experience in the Scillies and his nomadic existence was unlikely to impress a bank. My prospects were not much better. It was unlikely I could raise money for an unknown salvage project based on hearsay rather than fact. Apart from some cash in the bank, my only assets were a motorbike, a Land Rover and some diving gear. I didn't own the cottage.

Peter assured me that diving for scallops would provide a steady income. This seemed a rational plan, but we had no intention of giving up the dream of making a fortune through salvage. I was sure I could persuade the bank to give me an overdraft but my confidence ebbed as soon as I walked through the doors. The manager was friendly, and I explained my intention to buy a second-hand fishing boat that we would live aboard while using it to dive for scallops off the isle of Barra. I handed him the figures of the estimated costs and returns, assuring him that we would not take wages but live frugally on our earnings. Sitting in silence as he looked at the figures, I was unprepared for the next statement.

'I hear you've been undertaking salvage in the Firth of Forth.'

'Yes,' I replied. I looked at him curiously, wondering how he had found out, although I remembered that there had been an article about Dan in the local paper. I did not expand, as it had not been a particularly successful season and might act as quicksand to the project.

'Has it been successful?' he queried.

'I've made a living and bought some useful equipment,' I said. 'It's been good experience.'

'Are there any wrecks off the isle of Barra?'

I wondered if I should tell him. I knew I was a soft touch when people showed interest in salvage. It was like sneezing: I couldn't control it. I detailed some wrecks – the *Samuel Dexter*, the *Cyelse*, the *Seniority*, the *Maple Branch*, the *Jane*, the *Colonsay*, the *Baron Ardrossan*, the *Empire Homer* – with their various positions around the coast, the dates they had been lost, their size and tonnage, ending with the SS *Politician*, the wreck on which the film *Whisky Galore!* was based. My story of going to dive for scallops was quickly losing its credibility.

'The wrecks sound more interesting than the scallops,' he said, as he flicked through the pages of my old bank statements.

I had fallen into a trap of my own making and felt foolish. 'Yes,' I added sheepishly. 'I'm sure we'll have a look at them when we're out there.'

'Do you think they're worth salvaging?' he asked.

'They might be,' I answered warily – the damage had been done, so I was not going to add to it.

'The scallops are to be your main income?' he reiterated, as he placed my file back onto the desk.

'Yes,' I answered, adding, 'We'll look at the wrecks, if we get the time.'

During a long, uneasy pause, I was tempted to say some more positive words to convince him, but he looked up from the file and gave me a serious stare. *This is it*, I thought, *he's made a decision*.

'The bank will give you an overdraft. You must keep me regularly informed and send me a copy of the boat's papers.'

I felt like yelling it to the world. Standing up, I thanked him, being careful not to overdo it, but on leaving the bank I knew my dreams were a little closer. Now all I could think of was buying a boat. My mind quickly drifted back to the *Dewy Rose*.

The following day we drove up to Stonehaven to have another look at her. The town was quiet, the holiday season over. We found little change in the boat's condition – more green paint was peeling off the hull, and most of the fishing gear and all the safety equipment had been taken away. The compass brackets were twisted and broken and the compass was missing, along with all the other wheelhouse equipment. Below deck the whole boat smelt of damp and there were obvious signs of flooding. The engine still looked sound and I assumed it would run, although I was aware that might be wishful thinking.

Trying not to be too persistent, I made a final attempt to persuade the owner to sell by suggesting I looked in for a chat when in Aberdeen on other business. Now we had access to sufficient money we could make the owner, Mr Walker, an immediate cash offer. Entering the Fish House in Aberdeen, I saw Mr Walker beside the gutting table. He noticed me come in and slowly moved away from the table, took his overalls off and hung them up. I felt my heart pounding as he came across and beckoned me into his office. We already knew that a Bradford man had an option on the boat, which was due to expire on 1 October, but the harbour master had assured us that he no longer wanted it. Mr Walker agreed that if I paid cash today, he would give me a receipt stating that the money would be returned if the Bradford option was taken up. We had twelve days to wait.

2

Realising a Dream on Barra

When we heard the *Dewy Rose* was ours, we loaded the Land Rover with every type of tool we owned or could borrow, and with Yak sitting between us drove up to Stonehaven. Boarding the *Dewy Rose*, now our salvage ship, we were able to scrutinise her in more detail. The priority was to get her out to Barra and earn some money, leaving non-essential renovations to be carried out in the future, when we had more time and money.

I looked at Peter, giving him a shrug of my shoulders.

He looked back and said, 'Let's get started. It's October. It'll be a race to get her to Barra before the winter gales set in.'

We settled into a routine. The engine-room repairs were essential. Two good dry bunks were also a priority, along with replacing a section of planking on the boat's hull. A new wheelhouse had to be found and we hoped something might turn up.

After ten days the initial euphoria wore off, the enormity of the task becoming apparent. Peter suggested we threw all mental schedules away and took each day as it came. He was skilled at joinery and a neat painter, but was not mechanically minded and of little use in the engine room. The Kelvin K3 engine had been manufactured in Glasgow in 1956 and weighed about a ton and a half. Surrounding the engine was a clutter of various water pumps, generators and winch drives. I had to discover how they operated before I was able to make them serviceable. I soon became used to climbing down the oil-saturated wooden steps to the gloom of the engine room, where my clothes and hands became smeared with the black, pungent-smelling engine oil. It was not without a steep learning curve and the gift of an old engine manual that anything worthwhile was achieved.

We rarely ate on board, going to the baker for lunch and buying a fish or chicken supper in the evening. Yak, when fed up with watching Peter in the accommodation, would study me from the top of the engine-room hatch, wise enough not to join me below. The local people had been sceptical, but were curious to see what we were doing. They gave advice and watched our progress – or lack of it. The harbour fraternity, knowing we had a small budget, often helped us to obtain parts and materials, either free, second-hand or at the lowest prices.

Eventually I had the satisfaction of starting the boat's engine. It had three enormous cylinders. After opening the decompressors on each, screwing in various impulse mechanisms and connecting the electrical wires, I sat hunched beside the engine, my back against a fuel tank. Wiping the perspiration off my forehead with a clean rag I paused as I was left with the Herculean task of turning the cranking handle. With my feet on the timber frames either side of the boat I swung the handle backwards and forwards like a pendulum until I had gained enough momentum to swing it completely round. If it wasn't moving fast enough the pistons would bounce back on the compression, nearly pulling my arms off. If it turned there was a tremendous crack, as if the engine had exploded, when one cylinder fired, leading to a heart-stopping pause, a second crack, and a third crack, as the engine gradually built up speed. I bled the diesel system by opening valves and taps, counting to thirty before a large lever on top of the engine was thrown over. This activated the diesel system and stopped the petrol that she had been started on. The cracking noise changed to a *g-dump*, *g-dump*, *g-dump* and the whole boat vibrated in sympathy. I rushed on deck to check that the cooling water was being pumped through the engine, stumbling over Yak as she too watched the water being discharged from the side of the boat. We became engulfed in a large cloud of black smoke issuing from the exhaust, making an old guy ask, tongue in cheek, 'You're not a steamboat, are you?' After a few minutes it cleared and I returned to the engine room with a squeezy bottle full of oil to fulfil the golden rule of a Kelvin engine: if it moves, oil it.

Peter and I had become like brothers, knowing each other's weaknesses, strengths, annoying habits and the subjects or words that would cheer or irritate each other. Peter could charm the birds out of the trees and shoot them, as the saying goes, a trait which frequently worked to our advantage in getting equipment.

With the engine running and the refit of the accommodation adequate for the two of us, the final necessity was a new wheelhouse, as the original was so damaged it had had to be replaced. Neither of us wished to live our lives below deck: a nice light wheelhouse with a cooker, seats, stove and sufficient space to change into our diving suits was the ideal we were searching for. We discussed our predicament with the local fishermen. Within a few days they came back with the news that there was a wheelhouse being used as a garden shed in Peterhead. The owner was moving and would be delighted to sell it. But there was a catch. The garden had been blocked off by a new house – we would have to dismantle it to get it out through the remaining narrow alley and it had to be removed within the next three days. The wheelhouse was perfect, much bigger than the original, but still with room for a man to pass on either side of the deck.

We purchased it without much thought as to how we would take it apart and put it back together. If it had been built once,

surely we could build it again. However, when we went to dismantle it, we realised the mahogany-built structure was too well
assembled to survive such treatment without irreparable damage,
and in the end it would involve time and expense we could ill
afford. Worried, I went to fetch more tools from the Land Rover
and it was while there I noticed a crane working on a new house at
the end of the road. After a quick chat with the driver I arranged
to 'borrow' him and his crane to lift the wheelhouse out of the
garden during his lunchbreak. Unable to find a lorry at such short
notice we tried to lift it onto the back of the long-wheelbase
Land Rover. I was sure it could take the extra weight, but it was
not quite long enough; the wheelhouse was too wide. It would
stick out two feet either side!

We carefully placed the slings around the wheelhouse. As the
cottage obstructed the crane driver's view, the signal to hoist had
to be given through a relay of spectators. Peter and I guided it
clear of some shrubs before running round the cottage to watch
it swing sideways as the crane slewed. I had the dreadful thought
of what would happen if the slings slipped and it fell onto the
house. The wheelhouse, dangling on its long slings, looked as
though it might not clear the roof. Yak barked, and the house
owner looked horrified when the crane struggled to lift the jib
high enough.

The wheelhouse skimmed over the ridging to be safely guided
onto the back of the Land Rover. It fitted neatly with the predicted two-foot overhang either side but the excess weight on the
back made the front of the Land Rover point skywards.

A few miles north of Stonehaven we rounded some bends
before I slowed down, knowing there was a steep turn under
a railway bridge a quarter of a mile ahead. We found ourselves
having to stop behind a queue of traffic. My heart sank. Peter
jumped out with his usual enthusiasm, running forward to find
out the cause of the hold-up. Five minutes later he was back and

explained that there had been a car crash under the bridge. Two police cars were waiting for a breakdown truck to remove the car. Peter returned to watch the progress.

It was dark by the time Peter came back with a group of drivers from the cars behind us.

'Pull out and go to the head of the queue,' he shouted, his arms waving me forward to go ahead of the cars in front.

'What about traffic from the other way?'

'It's been stopped. Come on, we haven't got long.'

Peter ran ahead, waving me forward as the Land Rover with its wide load squeezed past the other cars in the queue. Creeping along with only my sidelights (the headlights didn't work), I saw one police car was ahead of the truck, the other at the side of the road, indicating the clear passage to follow.

'Drive on, drive on!' Peter shouted above the noise of the engine, as he jumped in beside me. 'They're going to escort us to the harbour.'

I followed the breakdown truck with its flashing light into Stonehaven, where the truck turned off into a garage. The police car continued to escort us until it stopped as we approached the harbour and waved us past.

The officer leaned out of the window. 'We're off to another incident. Good luck!'

As I parked the Land Rover alongside the boat I couldn't believe how fortunate we'd been.

The derrick creaked when the boat leaned over to take the weight of the wheelhouse and gently lift it off the Land Rover. An adjustment on the tackles by Peter placed it amidships on the aft end of the *Dewy Rose* and down she went onto the deck. We both clambered up the pier, walking backwards away from the boat to see how it looked. *Considerably better* was the joint opinion.

There were numerous smaller jobs to do – the list could have gone on forever – but there came a time, as in all ventures, when a decision had to be made. If we set sail too soon and broke down, people would consider us foolish; if we stayed much longer, the gales would blow and it might take months to get to Barra and we might run out of funds en route. So we made the decision: the motorbike was lifted into the hold, the hatch boards secured and the wooden dinghy placed upside down on top. The Land Rover was sold, the cash received allowing us to become owners of two wrecks lying off Barra. Peter had worked hard to persuade the owner of the wrecks, a ship breaker, to sell. The breaker had then waited patiently until we could produce the cash before concluding the sale. The deal gave us the legality we required for salvage work.

At last, we were ready to go.

We were sad to leave Stonehaven, having made many friends during our six-week stay. We had received much kindness and the only way to repay it was to succeed in our venture. We were proud of our little salvage boat and the work we had put into her. She had been a wreck, but we had brought her back to life, and in a week we should be tying up on Barra.

The boat went well, with only one major breakdown near Lady Rock off Mull on the west coast, which I was able to repair and we were back on our way. We passed inside Muldonach, a large mountain-like island, before following the navigational pillars on rocks and floating buoys used as channel markers on the way in towards Castlebay, the main settlement on Barra. Turning to starboard, Kisimul Castle appeared, standing proudly on a small rocky island at the head of the bay. The pier lay behind it, a small crowd of people appearing to be waiting; behind them was the 1,000-foot hill of Haeval, with a Madonna standing in a prominent position on its side. The bay had a fairy-tale quality.

We had arrived. This was it.

It was more impressive than I had anticipated, but I felt nervous at the thought of landing, curious to know how we would be received. I thought of Robert Louis Stevenson: 'To travel hopefully is a better thing than to arrive.' I would soon find out.

Our arrival coincided with the ferry from Oban, which explained the small crowd on the pier. 'The ferry's arrival is a social event,' explained Peter. 'That's why there're so many people.'

'Are we allowed to use the pier?' I asked.

'If we go round the back, we'll be well clear of the ferry.'

There was no shortage of people to take our ropes. Relieved to tie up, I wanted to express that relief and excitement to Peter, go over the trip, savour the achievement, but Peter was leaning out the wheelhouse window chatting happily. He was on his best form as questions were asked about the boat. She was being scrutinised in every detail.

I put the kettle on when Ken, the pier master, made his way through the crowd and talked to Peter about the boat's berth. He would allow us to stay on the pier as long as we kept clear of the landing point for the main ferry and Hector's small ferry from Vatersay, the next island to the south. I was half expecting an attractive young lady to appear, probably called Kay. I guessed from past experience that somewhere on the island, a girl – or girls – would be involved in some part of Peter's plans, but it had been nine months since he had left, a long time in any relationship. I looked around. Most of the people seemed to be much older than me.

When the ferry left, I kick-started the motorbike and the engine burst into life. Yak jumped on the fuel tank, and with Peter behind me we set off to have a look around the island. The circular road followed the coast, making a 14-mile round trip. Halfway round the island, on our return to Castlebay, Peter pointed out the sign to Northbay and Bruernish. It was here that Peter knew a crofter he hoped would rent us a house.

Time stood still as we sat in the crofter's home. Conversation was slow and relaxed. Peter spoke about his summer and asked about the crofter's year before finally coming to the subject of the house. I patted Yak and looked out of the window, enjoying the cup of tea. I was in no hurry to move. I could see small islands and inlets at Northbay, and not only was it beautiful but the water also looked crystal clear for diving. This was a pleasant change from muddy water on the east coast. I could easily live here, I thought, and couldn't wait to get into the water.

We left on the promise that a croft house would be available in a week or two; it had been uninhabited and needed cleared out. Continuing around the island, we stopped briefly to speak to a lobster fisherman Peter knew. I listened to him explaining to Peter that they were getting grants to buy new boats to increase their catching ability. Feeling a bit like a chauffeur – introduced, but only listening in to the conversation – I digested the new world that was opening up before me, trying to remember facts about lobster catches, or places that would be good to dive for scallops. The fisherman's dog looked aggressively at Yak, who huddled against me for reassurance. I stroked her gently and wondered if the dogs were related. She had my sympathy, as I knew how difficult some relations could be.

Barra had no conventional harbour. For the present, alongside the pier was convenient for access to the shore and when we moved into the cottage we could moor the boat in Northbay and go out to her with the dinghy. The weather was good and everything was going our way, as Peter knew of a small patch of scallops in the bay, which gave us an easy start. Going out with the dinghy, I trailed my hand over the side, letting the water run through my fingers. It had a different feel from the Forth, the clarity emphasising its purity. I longed to enter it but waited for Peter to dive first; he was going to check the density of the scallop beds to make sure he would be able to recover a worthwhile amount. When my turn came it was wonderful, the visibility changing little from the surface to the seabed. The colour of sea creatures was no longer grey. Anemones in particular looked like stunning flowers. The scallops lay scattered on the sand, and as I moved along crabs and small creatures scuttled out of my way. Broken bits of oarweed and kelp covered parts of the seabed, and empty scallop shells lay among the live ones. There were a few starfish, the main predators of the scallops. The starfish holds onto the shell with its tiny suckers and forces it open to allow its stomach to push out of its mouth and into the shell, dissolving the meat inside.

I had never experienced such good diving conditions. My fins kicked up small clouds of sand behind me that drifted off in the current; I concentrated on filling the sacks with big scallops, leaving the small ones to be harvested in future years. My two sacks were supported by a hook on a plastic five-gallon drum, used as a float. I could adjust the buoyancy of the float as I filled the sacks by taking the valve out of my mouth and injecting more air into the opening in it. With my air almost exhausted I had collected two full sacks, and filling the float with the remaining air I followed it to the surface.

Returning to the *Dewy Rose* after a second dive, we stitched up the tops of the seven sacks ready to be put on the ferry to Crofters Seafoods at Oban. That evening a small crowd of Peter's friends crushed together in the shelter and warmth of the wheelhouse while Peter fried scallops, along with a small piece of bacon, to make a tasty meal for us all that was washed down with cans of beer.

Peter liked a hectic social life and it appeared to be expanding in tandem with our income. On Saturday night we went 'on tour'. Leaving the pub after closing time with two half-bottles of whisky, a few cans of beer, two girls and a fisherman, we drove round in a van Peter had borrowed until we found a house well lit up and with cars outside it. After helping to push one of the cars out of the ditch we were shown into the main room. The bottles started to be passed round, as a bucket full of cuddie – small fish about six inches long that live around rocks and piers – were rolled in oatmeal and fried, the flesh peeling off the bone as we ate them. I felt I had hardly slept when I was woken by church bells the next morning, not in the best condition to go out diving!

Eating on the *Dewy Rose* became easy on three days every week; this was when the MacBrayne's ferry came in and we moved the *Dewy Rose* back a few feet to place her stern opposite the galley porthole of the ferry. Chatting to the cooks, we would exchange some scallops for three-course meals, passing the empty plates back and thanking the cook as he gave us mugs of coffee. We would miss this when we moved into the cottage, but there was no word that it was being cleared out, although we regularly phoned from the phone box. Peter realised it would never be done unless we put some pressure on and, in doing so, I discovered the phone required a polite touch, asking the operator for the person by name and always adding the word 'please' or they would not connect you. Time was of little importance on the island; the church and peace of mind were the rules of islanders' lives. Not that we attended the church.

Our initial euphoria began to ebb as the weather started to worsen. The wind rose and the *Dewy Rose* banged against the fenders, making it difficult to sleep. Our wetsuits wouldn't dry between dives – putting them on required handfuls of talcum powder to stop them sticking. But it wasn't the diving that was cold and uncomfortable, it was huddling in the open boat in a damp wetsuit, covered over with oilskins, waiting for Peter to complete his dive before I could enjoy the peace and quiet beneath the waves. Peter felt the same when he waited for me.

Soon winter truly set in, with severe gales stirring up the sea, making our work impossible in the wooden dinghy, and when using the *Dewy Rose* we found her uncontrollable in the rocky areas where the scallops lay. We did have some luck, however, when Peter purchased the remains of a wrecked puffer for a bottle of whisky.

The *Colonsay* had sunk twelve years earlier and its mast was still sticking out of the water. She had been built at the end of the war and sold to a small shipping company for taking general cargoes around the west coast. Sheltering from a storm in Castlebay, she dragged her anchor and ran aground before slipping off the rocks and sinking in 40 feet of water. It was to be our first salvage project. The site was so sheltered she could be worked in almost any weather conditions.

After we tied the *Dewy Rose* to the *Colonsay*'s mast, I looked over the side.

'The vis isn't as good,' I said.

'It'll be the water run-off from the land. We're opposite the entrance to a small sea loch,' Peter replied.

'It looks peaty.'

'Just imagine you're diving in whisky,' answered Peter.

As soon as I put my head underwater I could see the whole puffer: it was only about 80 feet long and 20 feet wide, but it was an impressive sight. It was the first time I had seen a complete ship on the seabed, and this was poor visibility for the island! The hull was covered by sea urchins, sponges and starfish, with small fish swimming among the weed. We swam down through the engine skylights to look at the compound steam engine, which had two large cylinders. The steam, after it expanded in one cylinder, was passed to the next, larger piston to do some more work before being condensed back to water in the steam condenser. The condenser had brass tubes, which were exactly what we were looking for and we quickly blasted them off and lifted them aboard, along with copper pipes and valves, recovering about half a ton of scrap. The *Dewy Rose* was doing what we had intended. The brief outing renewed our enthusiasm for salvage and gave us an unusual bonus, as most of Castlebay had heard the explosives go off. When we returned to the pier, there was a curious crowd waiting, giving us a ready market for the few scallops we had recovered from around the wreck.

'What did I tell you?' said Peter. 'Wait until you see the other wrecks.'

As the poor weather persisted the continual pounding of the boat against the pier caused the decks to leak and the sleepless nights were doing little for our tempers. We were desperate to get into the cottage. Peter said countless times he'd had enough and wasn't going to spend one more night on the boat. I felt the same.

So it was with relief that at last the day came when we could move in to the cottage. The rent was nominal, so off we went, full of promise, in an old car we had bought that shook and swerved in the wind as we travelled the five miles from Castlebay. The cottage lay 100 yards from the sea, the front door leading straight onto the road, with no fence around it or any sign of a garden; sheep grazed up to its walls. There was no water, electricity or toilet. Water had originally been taken from the roof, however the collecting barrel was missing, though opposite the house, in the grassland on the other side of the road, we found a standpipe. The typical croft kitchen-living room contained a sink, open fire and a few simple chairs with a small table. The cottage was incredible – compared with the boat, it was a palace, stationary and dry.

When there was a lull in the wind in the middle of January, we quickly moved the *Dewy Rose* from Castlebay to a mooring in Northbay that was close to the cottage. The morning after we moved her, the shipping forecast confirmed our fears: severe gale Force 9 for the Hebrides. I slipped out of the house to get a view of the boat: she was rolling and tossing, looking like a young girl enjoying a dance, swinging from one side to the other – one minute throwing her head right up, the next her bow down disappearing in a mist of spray. It was a beautiful sight. I felt proud of her, although that did nothing to lessen my concerns about the strength of her anchor and chain.

Unable to work because of the weather I had now grossly exceeded my overdraft and for four days we ate nothing but porridge: we boiled it, fried it, ate it cold, hot and tepid. Even Yak turned her nose up at it. The gales seemed to be relentless. Peter, undeterred by our present financial situation, made light of it. It was no real hardship, just disheartening. We both knew the weather would change and it was a time that brought us as close as we would ever be. Our neighbours were wonderful. When rumour got round that we were unable to work because of the weather, we were quickly provided with bread, eggs, milk, tea and butter. They would not let us refuse their help and this made us all the more resolute to go out, gale-force winds or not, to fend for ourselves. It was no use trying any salvage, as the wind was too strong; we would never have been able to hold the boat over a wreck, most being in dangerous places. But there was always a chance of getting some scallops in sheltered water. Peter undertook the diving, as I tried to keep the dinghy with her old Seagull outboard off the rocks and as close to his bubbles as possible. Yak would snap at the lumps of water as they came over the bow. If we'd had a rubber inflatable, we would have been able to work in much worse conditions. It was an item I had already added to my wish list, though in retrospect it would have been a better investment for the winter than the *Dewy Rose*.

When the good weather returned, it was fantastic to get out again in a calm sea and gather large quantities of scallops. Payments started to filter into my account. We dived early in the morning and then ate, taking at least a four-hour twenty-minute break before diving again. I was never sure about this rule, as it did not comply with the Admiralty Decompression Tables, but Peter assured me that it was safe. After the final dive, I enjoyed going to the beautiful sandy beaches on the west, walking along the coast or digging for an evening meal's worth of cockles on the strand. If it was low water and I was lucky with the timing, the De Havilland aircraft might land on the strand, the saltwater thrown up in a cloud by the wash of its propellers as it taxied to deposit its passengers at the edge of the beach.

While I was making the most of the freedom of Barra, Peter's mood began to change. After a few months there, he seemed to enjoy the social life less. I put it down to the change of company – some of the younger generation had left the island and most of those who remained were heavy drinkers, people he had previously been reluctant to socialise with. When we'd arrived on Barra, he'd said 'drink is like a disease', and with certain people on the island it definitely was.

Peter was often late and occasionally hadn't had any sleep. Undeterred, he would jump into the dinghy and we would set off to dive. During the twenty-minute run to the scallop beds Peter changed into his diving suit, which was awkward in the boat and not helped by Yak thinking it was some sort of game and climbing over Peter's semi-naked body. As we became more successful and earned more money, the length of his absences increased. It was only when he had spent most of his share of our earnings that he became eager to work again.

We made good money on the scallops, but it was like picking raspberries – no longer an exciting job, although diving in the clear water was always a pleasure. With Peter's enthusiasm for Barra on the wane, I was selfishly worried that other divers might come to Barra and clean out the wrecks before we had even made an attempt on them.

3

Island Salvage

Lulled into thinking the gales had passed for the winter and the calm weather would continue, we returned to working wrecks. We began with an old steam trawler before going to another vessel we owned, the *Seniority*. We had difficulty finding her until a local lobster fisherman offered to take us out in his boat. Although the wreck had been sunk for more than thirty years, it was still possible to see little droplets of oil float to the surface then spread out, showing all the colours of the rainbow in the bright sunlight, before they disappeared.

The *Seniority* was a 3,000-ton cargo ship, originally built in 1942 as the *Empire Boswell* for the Ministry of War Transport. On 7 November 1950, she had run aground at Leinish Point on Barra. Although refloated, she sank the next day off Bo Vich Chuan Rock.

Peter dived first with a buoy to mark the spot. I followed, reaching the seabed at about sixty-five feet. Parts of the ship lay on their side in the most fantastic visibility, some sections sitting on a sand and shell-grit seabed that contained broken shells from scallops and mussels. Large boilers were clearly visible, with long, eel-like ling swimming round the many holes in the wreckage and appearing unconcerned by my presence. I swam towards the engine room, where all the valuable metals lay. I could see part of the steam engine poking through the wreckage. Some of the brass had already been removed, but there was sufficient remaining to make it worth blasting free. I assumed the propeller and other easy items had been taken by Northern Shipbreaking, from whom we had purchased the wreck. The diving in Barra had already been stunning, but the clarity, sea creatures and weed, particularly kelp on this wreck, made for unbelievable underwater scenery. If all the wrecks on the islands were as clear as this, I never wanted to go back to the Firth of Forth.

When we returned and moored the *Dewy Rose* above the wreck, we took it in turns to dive as we lifted the brass. At times the *Dewy Rose* lay unmanned on the surface, Yak keeping watch on deck. I laid the explosives, as Peter gathered piles of scrap. It was like being paid to go on a diving holiday. Peter was as excited about the wreck as I was, so we worked our maximum times before both of us, time expired, lay sleeping on the seats in the wheelhouse, waiting for the time to pass before we could dive again. This went on for three days until the weather quickly started to deteriorate and I went to start the *Dewy Rose*'s engine, noticing water was pouring from a crack in one of the cylinder heads. We dropped our mooring and headed for the shelter of Northbay. With an overheating, unreliable engine and the increasing wind blowing us towards the shore, our hearts were in our mouths as we passed close along the rocky coast. On reaching the safety of Northbay, the wind had reached gale force. We checked the mooring was holding fast before realising that it was becoming marginal for taking the dinghy ashore. I abandoned work on the engine and Peter took the outboard from the wheelhouse and gradually started to pull the dinghy from astern up to the side of the boat.

We were rolling violently, getting soaked as we bent over the rail. One minute the dinghy was level with the deck, the next it was six feet below us.

'Now!' Peter yelled.

I jumped, clasping the outboard to my chest. A loud tearing sound came from my trousers as I landed in the dinghy. It was one of my few remaining pairs. Peter immediately slackened the rope, allowing it to drift astern. I fastened the outboard. It started first pull and I edged the dinghy back up to the *Dewy Rose* until Peter and Yak could jump aboard. Making little progress against the wind, we turned across it. So much water was coming on board that I tried to shield the outboard with my coat. We had no chance of going to our usual beach, so we were looking for an alternative when the wind gave an extra gust and the sea hit us like a rock. The outboard spluttered to a halt, and we each grabbed an oar, trying our hardest to row against the wind. Peter caught his foot in the baling bucket and disappeared over the back of the seat. We could do nothing but laugh.

For a few seconds all the pressure of the day evaporated: we were Peter and Alec again, two young tearaways. After coming to our senses it took us another ten minutes to travel twenty yards to the nearest shore – a long way from where we wanted to be. We jumped out and pulled the dinghy up a burn, well clear of any waves.

Even after the weather improved, Peter was still restless on Barra. When we were diving, he was casual, pushing his time under-water to the extent that I worried he would get the bends. We were making good money at the scallops, so he was reluctant to go back to salvage. I felt he was a bit like Mary Poppins – the wind had changed, and he was bored, wanting to move on. But for the first time since I had met him, he had been saving money.

After a long but successful day diving for scallops we were sitting at the wooden table in the cottage, the fire blazing with driftwood and coal as we warmed up after shedding our diving suits. This was always a relaxing time, comfortable in each other's company, neither of us wanting to move as we enjoyed the moment.

'I'm fed up,' Peter said out of the blue. 'I'm going to go to Spain for two weeks' holiday.'

I sat up in surprise. 'By yourself?' I asked.

'Yes, I need to get away from here and decide what I'm going to do.'

I couldn't help but feel let down – there were times when we could have shouted and screamed at each other with good reason, but never did; we were friends. The way he was talking, it was as if it was nothing to leave me in the lurch and head off on his own, but I knew there was no stopping him once he had made up his mind.

If I was honest, it wasn't really a surprise. Over the previous few weeks I had seen him change direction and had planned for it in my mind, was almost looking forward to a change. When Peter's moods were dark, they could bring me down with them.

But I wasn't ready for him to go just yet. The weather was becoming settled, I could see the chance to buy more diving equipment and a new rubber inflatable that should increase our earnings. I had hoped he would stay a little longer.

'When are you leaving?' I asked.

'I'll get the ferry tomorrow night.'

'Is it the diving you're fed up with?'

'Barra's not like the Isles of Scilly. It was always busy there, with tourists enjoying themselves,' he replied.

His past life was still a mystery to me. Whatever the reason for the holiday, I thought it would be positive: it might clear the air between us, so a decision could be made about our future partnership. Falling out was not the answer.

One morning, after Peter had been away for ten days, I saw my neighbour walking across to the cottage. He'd come to give me a message. 'Peter phoned,' he said. 'He's run out of money. Can you mail some to Glasgow, so he can get back?'

'Of course I will,' I replied.

We both laughed. It was all so typical of Peter.

When Peter came back from holiday, he looked well, with a suntan and a smile, bringing a bottle of wine and a bag full of fudge doughnuts and custard slices from the mainland to celebrate his return. His restlessness was more marked than ever, though, and on his second evening back he brought up the subject of the future.

'We'll have to stop diving for scallops at the beginning of the summer,' he said. 'They'll start spawning then.'

'What about salvage? We've the *Dewy Rose*,' I ventured.

'I was thinking of going back to work with Dan. I phoned him, he's bought a bigger boat.' He patted Yak while he waited for my reaction.

I looked at Yak. If Yak could vote, I thought to myself, she'd stay on Barra.

'I don't want to go back,' I said. 'We'd be throwing away the salvage here.'

Our limited wreck diving on Barra had been fantastic and I looked forward to working more island wrecks.

Peter interrupted my thoughts. 'I'll go back to the mainland as soon as Dan's new boat's ready.'

'Well, I'd rather take my chances here. What about the *Dewy Rose*?'

'I only need some diving gear and a compressor,' Peter said.

We looked at the accounts. Peter had contributed very little to the cost of the *Dewy Rose* or the wrecks, whereas I had sent nearly every spare penny to the bank. There were no arguments; it was settled in a few minutes. He took the old car, some new diving gear, one of our compressors and a wad of cash. I wondered what I would do with the *Dewy Rose*.

'That's it, then,' Peter said, getting up from his seat. 'We'll carry on here until our catches drop.' He looked relieved, and his old carefree happiness returned. 'How about going for a pint to celebrate?'

'Fine,' I said, amazed at how easy it had been. I should have known it would be like this. Peter was always decisive. In my heart of hearts I knew I would find a valuable wreck and it had to be good underwater visibility to make it worthwhile. Islands were the most likely places, as they were the least explored. My life had been standing still; I needed to follow those dreams.

Finally, Peter packed the car with diving gear and boarded the ferry. He stood at the rail with Yak. I felt sad, but relieved, as we waved to each other. I wondered if I would ever see them again.

When Peter was in Spain, I'd met Simon. He was working with a four-man diving team, headed by Chris, with another diver, Tony, and a marine archaeologist. They had recently come to Barra to salvage silver and gold from a Dutch East Indiaman lost in 1728 called the *Adelaar*. She had been outward bound from Middelburg to Java when she'd struck Greian Head on the west side of the island. They had discovered she had been salvaged not long after the loss by a Captain Roe, who had lowered men down in a barrel with holes cut for the arms, which were then bound in leather to keep the water out while a small visor was fitted for the diver to see. The barrel was regularly pulled to the surface to have bungs removed and air flushed through it using bellows. When Captain Roe had finished, all that remained on the wreck were several bronze guns, lead ingots and some smaller articles of value and interest which they quickly recovered. Chris had been working at Fair Isle during the previous summer on the Spanish Armada wreck of *El Gran Grifón* as an archaeological project. He had learned of Captain Roe's unsuccessful attempt to make major recoveries on the *Grifón* and then traced Roe's next trip to Barra only to find that Roe had been successful on the *Adelaar*. After hearing of their limited plans, my paranoia about competitors for wrecks or scallops eased and Simon and I became friends.

Simon lived in the Borders of Scotland and had trained as a journalist in Sunderland. He'd joined the diving team two years earlier to handle both their public relations and fundraising on archaeological wrecks. Having learned to dive the previous summer on the Armada wreck off Fair Isle, he and Tony had since been working commercial wrecks with Chris, on a similar basis to how Peter and I had worked for Dan. Simon casually mentioned that he might leave after the summer, although we didn't discuss it further. I knew this might be an opportunity, as he was a similar age and might be keen to join me on a salvage project.

Chris was the owner of Submarine Salvage Systems. He was tall, dark-haired and lithe. He had high ambitions and great determination, which was evident in the hours he worked. Seven years older than me, he knew his trade and had recovered valuable metals from wrecks around Scottish and Irish islands. There was little that he would not attempt and I learned fast from him, both about the legalities of salvage and the physical methods of dismantling wrecks. Speaking to Chris made me realise how little I knew about this type of salvage.

He asked me out to dive on the *Adelaar* – my first dive on an archaeological wreck. She was wrecked in a very exposed area, resulting in none of the wooden structure remaining. I was only able to see the lead ingots carried as cargo, cannons and bits of iron, all buried deeply in the boulder-strewn seabed. It was in beautiful clear water, where a large boulder had become wedged in the top of a gulley, adding to the stunning underwater scenery. When we came ashore, as if I had passed a test, Chris asked me to go to France with him and Simon to collect a fishing boat he wanted to buy. It was an old French crabber that he intended to use for salvage.

After tidying the cottage, I handed the keys back, paying off the rent due. Now the good weather had arrived, I could stay on the *Dewy Rose* again, having moved her into a perfectly sheltered inlet in Northbay.

Chris, Simon and I left Barra on the ferry to make our way to France, arriving at Douarnenez at five o'clock on a Friday afternoon, just a few minutes too late for the owner's agent to show us over the boat. We found the cheapest B&B and all three of us shared a room. On the way we had driven past a fairground in a nearby town, so we set off to enjoy ourselves over the weekend, wandering round the stalls, relaxed and sober, until we came to a 'goalie' stall. The stall had a small goal that was completely covered by a steel goalkeeper. The 'goalie' was made to swing like a pendulum, uncovering a small hole for a second that was just large enough to take a football. If you kicked three out of three balls into the goal you received a bottle of champagne. Simon could not resist the challenge. It took him a few shots to get the timing right, but then there was no stopping him. The ball went into the goal in quick succession, with a distressed stallholder not sure what to do. As Chris and I supplied the cash, the number of bottles beside us grew. We soon attracted a cheering crowd. The stallholder, concerned by the number of champagne bottles he was giving away, found a good English speaker, who asked Simon if he was a professional football player. Simon would have loved to have said 'yes', just for the accolade, but he shook his head and was allowed to continue. We knew it could not go on for much longer, and indeed it ended abruptly, the stallholder taking his ball back and closing up. Walking to the car carrying the boxes of champagne, with one bottle already open, we were asked to a party by one of the spectators.

Early on Monday morning we had a look at the fishing boat

before the agent appeared with two mechanics to start the Caterpillar engine. I saw Chris look at them with concern. *Why were they required? What was wrong with a starter button?*

'Accommodation won't cost us anything now,' said Chris, as he removed the padlock that secured the hatch leading below. I winced as a damp smell emerged. The messy and stinking cabin was all too reminiscent of my first visit to the *Dewy Rose*. Chris reminded the agent that the boat had to run and that he had previously stipulated that it must have a trawl winch, though there was none aboard. The agent arranged for us to see a winch that afternoon in a nearby estuary, but it turned out to be attached to a derelict trawler that submerged at high water.

Further inspection of Chris's boat revealed that marine worms had devastated the ship's frames and planks on the inside of the hull, where she had tanks for carrying live crabs. These were free of antifoul, as it would have killed the shellfish. I was shocked by its condition, frightened to rest my feet on the frames in case I put my foot through the bottom. This damage had not been visible when we looked at the outside of the hull.

The next day, Chris was still undecided. We stood on the deck looking at places where a new winch could be fastened until our silence was shattered by the engine starting. It had taken the two mechanics a day and a half. The noise was like rattling a biscuit tin full of stones. That evening we treated ourselves to a meal in a restaurant, as Chris wanted to discuss the situation and come to a conclusion. He was disappointed, but not crestfallen.

The following day we left the boat, returning to the B&B. We waited for the agent to come up with an alternative, but nothing was found and we returned to the UK.

Back on Barra, Chris offered to buy the *Dewy Rose*. She was in the right place for him and cheaper than the French boat, but I knew she was neither the size nor the sort of quality he had originally been looking for. With some feelings of regret I agreed to the sale.

I had thoroughly enjoyed myself on the trip with Chris and Simon, during which they told me about their recoveries from wrecks, Chris being an expert in dismantling with explosives, but listening to them also highlighted how little I knew. In terms of explosives, my only experience before working with Dan had been as part of a team blasting a tunnel through rock for a hydro-electric project, only taking the job to enable me to gain experience and to be able to obtain an explosive permit. I knew I could build up some more cash for a better boat while continuing to dive for scallops with a hired boatman, and reckoned I might earn a good living if I managed to find another partner. I was not going to make a fortune on the wrecks I knew of on Barra.

A few days later Chris asked if I would sell him the wrecks that Peter and I had bought on the island. Pondering his offer, I realised that without a proper boat and my lack of experience there was little point in keeping the ownership. The wrecks could be cleaned out at any time by someone else as the title gave little protection from a determined salvor. Plus, we only owned three. It would have been foolish not to sell.

With the cash in the bank from the sale of the *Dewy Rose*, I already felt six inches taller.

'D' you want to join us for a while?' asked Chris.

'What are you going to do?' I asked.

'We'll go round the islands, picking up scrap from wrecks we've previously worked with the inflatable. See if we can find some others.'

'OK,' I said, knowing it was a loose word-of-mouth deal, but it would be good experience, which I needed and would solve my accommodation problem. I would be able to stay on the *Dewy Rose*.

'When do we leave?'
'Tomorrow.'

Chris's large compressor and all his diving gear were lifted aboard the *Dewy Rose*. He left his heavily pregnant wife Jean in their house on Barra while the three of us set off, Simon looking forward to getting back to Coll and Tiree, where they had worked the previous autumn and winter before coming to Barra.

Our first stop was the wreck of the *Tapti* on Coll. We manoeuvred ourselves inside a reef to moor over the wreck that again lay in fantastic underwater visibility – that is until Chris started blasting, after which he insisted that one of us, usually him, should descend into the black water immediately to feel around and hopefully report that the blast had been successful. Any brass or copper was quickly secured to a rope and lifted aboard. This was repeated until all the easy bits had been recovered. As darkness fell each day we stopped diving and sorted the scrap on deck. Any steel attached to the brass was cut off with an acetylene torch or using a sledgehammer on the back of the blade of an axe. The steel was thrown back to the wreck, while the brass was lowered into the hold, which was gradually filling up. With the weather good and a favourable forecast, we remained moored over the wreck throughout the nights in readiness for an early start. Our meals rarely varied, being restricted to eggs, bacon, sausages, packets of cereal and tins. Simon's particular favourite was Carnation condensed milk, which he would drink straight from the can!

Work continued until Chris decided it was not worth staying any longer: a difficult decision to make, as the copper and brass never ran out – it just became harder and harder to extract, which meant we were taking more time.

The wreck I learned the most from was the *Hurlford*, a 444-ton steam collier lost in 1917 in Gunna Sound, the stretch of water that runs between Coll and Tiree. We stopped close to a navigation buoy that marked a submerged reef and prepared the inflatable. Chris dived first, straight down the buoy's mooring chain, and within five minutes found the wreck. He surfaced to take the end of a heavy rope, then swam back and secured it. The rope was winched in, pulling the *Dewy Rose* over the wreck.

'Nobody's been here before us,' exclaimed Chris, as he made up explosives on the deck. 'Everything's on it!'

I discreetly watched him as he lay the small charges on the wreck, all linked together by an explosive cord called Cordtex. I worked around him, lifting loose bits of brass out of the engine room.

Chris had explained his reasons for where he was placing the explosives. It looked as if he was blasting off material of no value, but it was just that some of the brass he was blowing was not even visible; it lay encased in a cast-iron housing, or clamped down under a thick steel backing plate. The visibility was good but I knew from experience to take a good look and remember some large objects to use as markers – after the blast we would never have the same visibility. When the explosives were laid we slackened our mooring, paying out an electric detonator cable at the same time, making sure the *Dewy Rose* was sufficiently clear of the shock wave that would result. The blast was to be set off using the boat's batteries.

I had briefly touched the terminals when Chris shouted down to the engine room, then I felt a heavy thump on the bottom of the boat before rushing up on deck to see dirty water well up on the surface, carrying large lumps of kelp. We winched back on the mooring, looking down at the disturbed water, wondering how well the explosives had worked. Taking a short break for something to eat, we waited for the current to clear the worst of the dirty water.

Returning to the wreck I could see pieces of brass shining everywhere like stars in among a dark cloudy layer that hovered over the seabed. In places it rose like whispers of smoke before being caught in the current and whisked away. I looked for large objects, like a boiler and part of a broken ship's side, to get my bearings, searching for the bits Chris wanted to lift first. Simon and I took it in turns to hook on the scrap, as Chris worked the winch. We lifted the brass steam condenser, gunmetal bearings, air pumps and general fittings before heading for Barra.

On the trip back I looked into the hold, trying to place the original position of the bits Chris had blasted off. If only I'd had Chris's experience when I was working with Peter! Having an engineering mind, I knew I could learn Chris's method very quickly. It was as if all the books I had read on salvage and steam engines now made sense.

Tony joined us for a trip to visit the wrecks on the islands south of Barra. He was the wit of the group, a true cockney. Apart from a sharp sense of humour, he could turn his hand to any trade – a real Mr Fix It.

Sailing south down the island chain, a local fisherman in his small lobster boat led us to various sites where we found the remains of wrecks but nothing worthwhile until we went round to the south of the island of Sandray. Fishermen, I was learning, were a great asset to us, as they knew where the wrecks lay, either by catching their nets on them if the ground was suitable for trawling or from hauling rust-coloured lobsters off them if they lay close in to the coast.

Chris had a quick look at the *Baron Ardrossan*, a 4,000-ton ship built in 1932 that had run aground in 1940, before we moved into an inlet where the larger *Empire Homer* lay. The *Empire Homer*, at 7,000 tons, was built in 1941 and had run aground in 1942. Copper and brass were immediately visible, which gave us all a boost. The inlet provided good shelter for the *Dewy Rose* from the Atlantic swell, and Tony and Chris prepared to blast while Simon and I lifted loose bits of brass, taking it in turns to dive and operate the winch.

The day passed quickly, but the swell started to sweep into the inlet as the light faded. Chris didn't want to leave – 'We'll get one more bit,' he said. 'Just one more bit . . .' Then it became too dark to work and our diving times were used up.

I couldn't have known that there was trouble coming. With the engine running, we released the aft mooring. But as the light wind caught her, the boat quickly swung round, drifting over the bow mooring rope, which caught in her propeller. We lay there, moored to the wreck by our propeller. I shut off the engine immediately. This was serious. We were drifting over the shallow part of the wreck and the rocks with no control of the boat.

Tony jumped into the water to free the rope, cutting it with a hacksaw before attempting to untangle it. Meanwhile Chris was shouting orders. 'Get the inflatable in the water! Quickly, quickly! Come on, Sy, put your back into it!'

Simon smiled at me, as the inflatable slipped over the side. Chris jumped into it to manoeuvre it round in an attempt to push us off the rocks. I felt the *Dewy Rose* bumping on the bottom.

'Prop's free!' shouted Tony, as he surfaced.

I quickly nipped down to the engine room and fired up the engine. High on adrenalin, we could see the kelp waving where it broke the surface and the outline of the rocks beneath us as they were lit up by the deck lights. Chris pushed her backwards and sideways with the inflatable to find a gap between the rocks, while the boat continued to lightly strike the bottom until she found a gap and slid into deeper water. I put her in gear to drive her clear.

On the way back we huddled in the warmth of the wheel-house, steam pouring off our damp bodies as we talked over the day, revelling in the exhaustion from the hard physical labour. There was a feeling of tremendous achievement, with the added piquancy of danger safely passed. Glancing out of the wheelhouse window we could see the hold and deck were filled with scrap. We looked with satisfaction.

Life was good for all of us.

Chris took a few days off to spend with his wife and new baby – this was the first time I had seen him stop. He worked incredibly long hours and had a dogged will to get things done. I had never met someone so determined. Simon and I were tasked with preparing the *Dewy Rose* for another trip. Our next island would be Fair Isle.

I was never sure whether Simon considered the *Dewy Rose* any better than the boat we had seen in France, but this was forgotten by him at the prospect of returning to Fair Isle. As the boat had minimal washing facilities, we threw the old sheets out and replaced them with clean white linen trestle tablecloths that we bought at a charity sale for a few pennies. With everything tidy and the diving gear aboard, the last item loaded was an old steel water bowser that took up much of the deck space. Initially curious to know how it would be used, I later wished we had left it on Barra.

Chris was happy at the wheel of the *Dewy Rose*. His usual tension had gone and he appeared completely relaxed. It was a mood I had rarely seen.

This was my first trip to Fair Isle and I was unsure of what to expect. It was a remote island, an eighth of the size of Barra. I had no concept of what the population of fifty-five would be like.

The shipping forecast was reasonable, but as we moved away from the shelter of the Hebrides the rolling motion increased. I took a spell at the wheel, meanwhile Simon fiddled with the radio on the bulkhead in the wheelhouse, attempting to improve the reception. This varied due to its inadequate aerial and the direction the boat was facing. Simon liked to listen to sport: any sport would do, but catching up with the football news was his top priority. Finding nothing to interest him, he went towards the accommodation. 'I think I'll get some sleep to pass the time,' he said, leaving me at the wheel.

Cape Wrath, at the north-west of the Scottish mainland, was starting to disappear behind us when Simon came up from the accommodation, pulled a jersey over his T-shirt and thrust his multicoloured bobble hat on his head. Coming into the wheel-house to take a spell on the wheel, he rubbed his glasses with an old rag to try and get them clean, muttering that they were like lavatory windows before putting them back on. He was always lightening the mood with his quirky ways of seeing the world.

Chris and I moved out of the wheelhouse to the hatch and sat on either side of it, enjoying some rest in the warm rays of the sun. I reflected on my time with Peter; we had been playing at it. Our recent recoveries were down to Chris's skill, enthusiasm and experience, and I knew I was learning. After this trip I thought I might set out on my own again.

Half an hour passed before I noticed the boat's course weaving before fixing on a position that seemed different from the one we had started on.

'Has the course changed?' asked Chris.

'I think so,' I replied, slightly surprised, as I had marked it on the chart.

We both went aft to the wheelhouse to find Simon with his ear held against the radio.

'What's your course?' asked Chris.

Simon, totally engrossed in the radio, was upset by the distraction. 'I've shifted it a few degrees.'

'Why?' asked Chris.

'It's the only way I could get reception to hear the end of the women's Wimbledon final. It's been getting fainter and fainter since we left the coast. Billie Jean King's beating Evonne Goolagong!'

We shook our heads as we returned to our seats on the hatch. We would have to change the course a few degrees when the match ended.

Arriving at the west side of Fair Isle with most of the day ahead of us, Chris was keen to dive on the wreck of the *Canadia*, which lay below the cliffs near Malcolm's Head. He had searched for and found the propeller when he had been working on the wreck of *El Gran Grifón* the previous summer but had lacked the equipment to lift it. The prospect of recovering this valuable object was uppermost in Chris's mind – one reason for buying a boat – but he was worried that someone else had lifted it. It was open season for wreck salvage in remote areas, with the Receiver of Wrecks – a branch of Customs – finding them troublesome to deal with; they normally received 7.5 per cent of the wreck value after a year and a day if the ship or cargo owner made no claim on the recoveries. The salvor would receive the rest. If the ship owner claimed the recoveries, he had to negotiate with the salvor.

The *Canadia* was a Danish steamer of 5,000 tons, lost on 12 March 1915. She had been captured by the British and was on her way to Kirkwall for inspection when she went ashore during the night on Heely Stack, a large rock pillar. The crew clambered onto land and were rescued by a breeches buoy the following morning. Her cargo was cotton and flour. The cotton bales had floated, allowing the islanders to tow them around to Kirki Geo to be recovered, along with some of the flour, which was prized by the islanders who grew a much coarser rye grain.

The bronze propeller lay in a narrow gulley beneath Heely Stack, where it had been forced down by the weight of the ship as she was pounded by the heavy seas. The broken remains of a brass steam condenser lay in a deeper gulley further off the rocks. This consisted of a bundle of brass tubes weighing several tons. Chris knew exactly what he wanted and how he was going to get it. I took the *Dewy Rose* close into the cliffs as he put on his diving gear. He knew the site well and pointed to a submerged reef between us and the wreck. I was to keep seaward of it when he swam in.

Simon and I chatted, keeping an eye on Chris as I nudged the *Dewy Rose* back into position each time she drifted off.

'I've told Chris I'm leaving when we've finished this summer,' Simon said.

'I'm only here for the trip,' I replied. 'I'm not sure what I'll do, but I'd like to stay in salvage.'

We were at the limit of being able to see Chris's bubbles as we talked, manoeuvring the boat closer to keep them in sight.

'It'd be nice to come back to Fair Isle next summer and search for wrecks around the island. Loads are recorded and there'd be some lost that no one knows about,' said Simon. 'After that I'd go back to a "normal" job as a journalist,' he added with a smile.

'It would be fun,' I said, 'but Chris is after the best scrap.'

'There must be other valuable wrecks,' ventured Simon, before being distracted by Chris waving his hand above the water and swimming back to the boat.

I took the *Dewy Rose* closer to the reef. Simon threw Chris a rope and hauled him alongside. He took out his mouthpiece to shout. 'The prop's still there, but it looks firmly wedged in the gulley.'

'The same as last summer,' said Simon.

'We'll get it this time,' replied Chris.

I went back into the wheelhouse to take the boat well clear of the rocks, while Chris decided what to do.

'Let's go round to North Haven and lie at the pier tonight. We'll get the boat ready for lifting tomorrow,' he said as he pointed northwards.

I opened the throttle and the boat pressed forward. I felt myself relaxing as we distanced ourselves from the cliffs. Fair Isle lacked the sandy beaches of Barra: it was all cliffs, stacks and reefs, a rugged, dangerous but attractive coast on this uninhabited side of the island. The air was full of seabirds, particularly fulmars, swooping down before gliding up with the moving air current. Cliff ledges were packed with them, lined up like supporters on a football terrace, the rocks beneath them white with excrement.

I thought of Simon's plans. I still had much to learn about salvage, but Chris had taught me the basics: all we needed was a good wreck. Fair Isle was a great place to start.

Rounding the North Lighthouse, we entered the Haven, leading to the pier. It was open to winds from the north and not a safe place to leave the boat unattended. Ahead of us, on the only sandy beach, the island mailboat *Good Shepherd 2* was hauled up into a shed on a slipway. We tied up on an open pier with a small hand crane on the end that could lift barrels of fuel or other smaller items. I liked what I saw and our arrival filled me with a deep satisfaction.

Of the three of us, Simon was the best cook, but our stores had been planned to avoid most cooking. We settled down to heated tinned chilli con carne, baked beans and instant potatoes followed by Ambrosia creamed rice with tinned fruit and custard, the pudding eaten straight from the cans. Interrupting our evening meal, two islanders arrived to see Chris and Simon. They were close to our age – a change from Barra, where everyone seemed so old.

Chris and Simon knew them well. I was the stranger, not sure what to expect, and my guilty feeling persisted, instigated by Simon – that we were 'pirating' the island of its valuable ship-wrecks. Simon was very much at home, asking after various people and explaining that we had already been to see if the *Canadia's* propeller was still there. They laughed – little happened around the island without someone noticing. Jerry, who had the nearest croft to the wreck, had seen the boat approach the cliffs and guessed who was on her.

When the islanders left, Simon said, 'When we get some time ashore, we'll go up to the bird observatory, as there may be some talent there.'

'Talent' was a general term Simon used to describe any girl. The observatory was the nearest building to the pier.

'It would be good to see the island,' I said. I felt fit, clean and confident to go ashore, as I had bought some new clothes, the old ones going straight to the engine room as rags. 'Will it take long to see the rest of the island?'

'The island's only three and a quarter miles long, one and a half miles wide. Everything's within walking distance,' said Simon.

'A shop?' I asked.

'Yes, Stackhoull Stores. It's mostly food.'

It turned out there was no chance of a look around. Chris had left his best inflatable boat with Tony on Barra and that evening we prepared the older one. The neoprene material had worn so thin in places that the escaping air left bubbles on the surface after we'd wiped it with solvent. We glued on some patches before assembling it ready for the morning. I looked at Simon, who shook his head in acknowledgement that this was probably not going to do the job; we would have been wiser to have taken our time and glued more patches on. Chris, who was always single-minded, determined to quickly complete the work, was satisfied with it and we moved on to prepare the diving gear.

The following morning we set off for the *Canadia*, eating our breakfast as we went. Chris and Simon took the inflatable, transporting the load of explosives into the wreck to blast the propeller off the shaft and cut the four blades off. This would make it light enough for the *Dewy Rose* to lift in five separate parts. I stayed on the boat, waiting until after the blast, when they returned to wait for the underwater visibility to clear. Half an hour later Chris took the inflatable in for a look and was disappointed with the result. I moved the *Dewy Rose* further offshore, as the wind started to freshen, bringing with it an increased swell, making it unlikely that we would dive again that day.

'You take the *Dewy Rose* back to the pier, Alec. Sy and I will take the inflatable into Hestigeo.' It was an inlet with a small concrete pier that lay close to the wreck; they intended to pull the inflatable out of the water and secure it with all the diving gear. This allowed us to leave the *Dewy Rose* at North Haven for a few days while we walked to the inflatable each day, using it to complete the blasting and lift any small bits. When the propeller was off and we had piles of brass stacked up on the seabed, the *Dewy Rose* would be brought back to lift it all.

Working from the inflatable, I became familiar with the reefs around the wreck site. We found ourselves working in rough conditions, with water breaking around us, which I would never have considered possible on our arrival. When it was my turn to go down to cut the brass tail-shaft liner off the propeller shaft, I had to swim down to the shaft, squeeze a small quantity of explosives into the crack between the brass liner and the shaft, connect a detonator with wires to the surface, swim back up, grasp the grab rope along the side of the inflatable and haul most of myself out of the water as Chris set off the detonator. Quickly diving down again, I placed another charge as I worked my way along the liner.

With the weather deteriorating, I climbed into the inflatable for the final time, but when shifting my right hand from the outside grab rope to the inside of the boat I felt my hand go straight through the rotten material. Throwing my bottles off, I gripped the torn edges together, trying to reduce the amount of air coming out.

'Chris!' I shouted. 'There's a hole in the boat!'

'Where? Where?'

'Under my hand,' I said.

'Let's see, let's see,' he shouted, tearing my hand off.

There was a loud exhalation of air. 'Shit!' shouted Chris.

It was a long time since the five individual air compartments of the old inflatable had acted independently. The boat began to lose its shape. Gripping the hole again, I started the outboard with the other hand while Simon cast off the mooring and we headed for Hestigeo, Chris endeavouring to work the foot pump. He looked around wildly to see how he could lighten the boat, and Simon caught his eye.

'Sy, you'll have to swim,' he said. So over the side went Simon. He was in his wetsuit and floated feet up to show his disapproval of the hasty decision. I reduced the outboard speed, as the shape of the boat continued to collapse. By this time it was like a flat raft, wallowing in the heavy swell, as we moved out from the shelter of the stacks. Motoring close to the cliffs, I knew that we needed to jump out before the boat touched them or we would be tangled with our equipment. I looked at Chris.

'Do you think we'll make it?'

'We've got to,' he said. 'Get as close to the rocks as you can. It'll shorten the distance.'

I nodded, not too sure about it, and shifted my position to be able to jump clear of all the ropes, explosives and other bits in the boat. Simon was swimming behind us, relaxed and easy in the turbulent sea, keeping well clear of the rocks. He was no worry.

At last we turned in towards Hestigeo and gained some shelter. Arriving at the pier with the boat awash, we quickly hauled the outboard, fuel tanks and diving bottles onto the safety of the concrete platform. Simon then dragged the airless inflatable to the stony beach at the end of the geo.

Repairs were quickly carried out but the weather soon worsened, giving us a few days off. With the wind from the southwest, the *Dewy Rose* could be safely left at North Haven, allowing us time to explore the island.

Every Saturday afternoon the Fair Islanders played a football match against the team from the bird observatory. The match was known as Fair Isle versus the Rest of the World. The previous summer

Simon had played for the Fair Isle team, for which he was a useful addition. I was asked to play for the Rest of the World. This team was made up of anyone they could persuade to play. Some played with binoculars strategically placed on the side lines – it was not unknown for the 'bird boys' – as Simon referred to both sexes of birdwatchers – to rush off if a rare migrating bird appeared. The only unwritten rule was that the match continued until the Fair Islanders won. They had some good players, particularly Jimmy, one of the younger mailboat crew, who interacted with Simon to score most of the goals. The Rest of the World was thoroughly beaten well before full time was called.

Bad weather had set Chris's schedule back, but when we did get out the *Dewy Rose* quickly lifted the piles of scrap we had gathered, along with three of the blasted propeller blades. A remaining blade lay jammed beneath the rock, with only a small knuckle of brass visible on which to secure a sling. As the *Dewy Rose*'s winch was unable to lift it, and the swell was increasing with the change of tide, she was anchored clear of the wreck before the bowser, loaded aboard in Barra, now came into its own. Dark green and oval-shaped, about eight feet wide by eight feet long, it had been used by the army as a water tank when mounted on a small truck. Chris had welded a three-inch-diameter tube right through the tank, from top to bottom, intending to thread a wire through the tube connected to a winch at the top and the object to be lifted at the bottom. This used the bowser's seven-ton buoyancy to lift any object below that weight.

The bowser thumped the side of the boat as we lifted it into the water and rolled wildly as it was pulled by the inflatable to the site of the jammed propeller. The winch wire through the tube was attached to the propeller blade 20 feet below. It was like a barrage balloon on a windy day, trying to lift the lorry it was attached to. It swung wildly with every wave, jerking to a halt when it was pulled up short by the anchored wire, changing direction to swing back the other way. Simon and I were expected to climb onto it and work the hand winch attached at the top. Neither of us managed to cling on in our diving gear, being repeatedly thrown back into the water. We swam back to the inflatable, taking off our gear before trying again.

This time, balanced on top, we worked the winch. It had a lever similar to a Laurel and Hardy bogie that ran along a railway track. We duly swung this up and down, and as the wire tightened the motion of the bowser gradually reduced, and instead of bobbing and swinging on the waves, the waves swept over us. Apart from the water, it was like being on a fairground ride. Chris, nearby in the inflatable, urged us on. Simon and I looked at each other, a smile passing between us as we continued to winch the wire in. The bowser was pulled deeper and deeper in the water, the remains of the propeller failing to break out of its rocky grave in spite of the lift transferred to it. With just two feet of the bowser sticking out of the water during the passing of the shallowest wave, we stopped and clung on to the winch, the bowser at times being completely submerged by the waves, along with us and the winch. I looked towards Chris in the inflatable, and he nodded and waved, indicating we should continue. Eventually, when the bowser was completely underwater, it became pointless to winch it further, so we jumped off and swam to the inflatable.

'Let's sit and watch,' said Chris. 'The propeller should wiggle its way out of the rock with all that lift.'

'We'll worry it,' said Simon with a smile on his face. I smiled back. The thrill and danger of the work was like the high of taking a drug.

We waited and waited, bobbing up and down in the inflatable, watching the waves rush over the bowser and crash against Heely Stack. There was no sign of the bowser rising, which would indicate that the propeller blade was loosening. Eventually, when Chris had finally accepted the propeller was not going to budge,

Simon and I returned to release the wire. This was a nightmare operation, going through the whole process in reverse. When the bowser was free, we pulled it back to the anchored *Dewy Rose* and lifted it aboard, unable to avoid it crashing against the side of the boat. On the way back to the pier I could see by the bilges that the *Dewy Rose* had a leak. It wasn't bad, but it was new.

'We've got a bit of a leak, Chris,' I said, as I came out of the engine-room hatch.

'Wait until we get to the pier and we'll have a look,' Chris replied. 'It'll be where the bowser's hit us.'

Back at the pier Chris dived in to investigate; the *Dewy Rose* was well down in the water with the load of brass, some loose bits were even stored in the accommodation to leave room for the remaining propeller blade in the hold.

'It looks as though some of the caulking's out,' he shouted from the water.

'I'll pass down an old bit of rope to tap in to replace it, shall I?' I asked.

'I've stuck some weed from the pier in for the moment,' replied Chris. 'We'll wait until morning to see if it's stopped. We'll know if that's the place'

The following morning I went down to the engine-room to start the engine. 'That's the leak stopped,' I called up. 'D' you want me to dive to fill it properly?'

'I'll do it,' said Simon, keen to have a search under the pier. It's amazing the bottles and other interesting souvenirs that get dropped over the side.

'We've not got time,' replied Chris. 'Don't worry. The water pressure will hold it in.'

Simon looked at me and smiled before saying in a confidential voice, 'Just like the inflatable.'

We left the pier for one more try; we would catch the ebb tide that drew the swell off the west side, but it had increased overnight. Lying off the wreck in the safety of deep water, we watched the waves as they washed over the reef. At odd moments when the sea appeared calmer, we ventured towards it, only to be chased back by a giant breaking wave. Simon and I glanced at each other – we knew how determined Chris could be. I went forward to the accommodation, filled my little leather suitcase with my cheque book, underwater camera, diary, letters, passport and wallet. Taking it on deck, I placed it on top of the winch.

'What's that for?' asked Chris.

'It's in case we break up on the reef; it contains all my valuable possessions.'

Simon made out to look as though he was going to do the same.

'Okay, okay. We'll go back to North Haven,' said Chris. 'The tide will change soon and it'll worsen.' Simon and I looked at each other like two conniving children.

Chris went up to the bird observatory to phone his wife and returned with the news that we would sail for Barra. Setting off in the last of the evening light, we passed the northern islands of Orkney with the old engine chugging away rhythmically. I followed the routine checks on the engine, not expecting any problems. The hours passed and then a day. We approached Cape Wrath, hoping to clear it by about five miles to avoid the worst of the current. Visibility was good, with a windless swell slowly rising and falling as we neared the Cape.

The first we knew of a problem was a violent banging and vibration. I hurriedly shut off the engine. There was silence. The boat began wallowing in the long shallow swell. Chris and I started to take off the inspection plates; the oily bolts were easily slackened, allowing the access doors to be lifted clear. What we saw was not good. The crankshaft was broken within one of the big end bearings. We came up from the hot stale air of the engine-room and looked at the shore. It would have been possible to tow the *Dewy Rose* for a few miles using the outboard motor, but we did not have enough fuel. We had to try some sort of repair. We weren't concerned about what we did to the engine because it was scrap now and wouldn't be worth the cost of repair. Blanking off the cylinder with the broken crank beneath it, we cut the connecting rod, using the lower half as a clamp on the broken shaft. If the engine could run on two cylinders, we might be able to crawl towards the coast.

The engine started with tremendous vibration and the kind of noise an engineer would find as irritating as an out-of-tune orchestra would sound to a musician. We ran it as slowly as possible; the nearest safe haven for anchoring was Loch Eriboll. Our chart would get us close and then we would have to guess the final course to a faintly marked pier. After several hours the relief at seeing it was overwhelming. The ancient stone pier looked neglected, with no indication that any boats were using it and no signs forbidding its use. We managed to get the full length of the *Dewy Rose* alongside before she touched the bottom. I was hopeful that this could be our base until Chris made a decision. He wandered ashore to find a telephone, while Simon and I checked the ropes to make sure the boat would lean against the pier when the tide went out, leaving her high and dry.

Chris had told Simon and me on Fair Isle that he would no longer need us when winter approached; diving on wrecks became uneconomic and he was considering going back to Tiree, or returning to the south-west of Ireland, where his wife had a share in a business. Financially, in the last few months, Chris had made the value of the *Dewy Rose* many times over. We would have to wait and find out whether he wanted to replace the engine or whether he would, because of this setback, call it a day. Neither of us wanted to leave him in trouble, but there was little we could do other than tidy up.

The following morning we were still in our sleeping bags when

Simon opened the hatch to see a man looking down at us from the pier. Simon tidied himself up and went out to meet him. Peter had relied on almost a military type of politeness, with a fairly firm authoritative manner. I had always been surprised by how well it had worked. In contrast, Simon was gently talkative, conciliatory and understanding, relying on winning people over by convincing them that we would behave. This tack would be difficult, as we must have looked like nautical tramps. Initially, the visitor had his hands on his hips, looking quite confrontational, but after a few minutes he appeared to relax. When Simon glanced towards us, Chris and I went to meet him.

'My boss won't be up here for a couple of months. If you don't make a mess, you can stay to unload and repair your boat,' the man said.

Chris arranged for Tony to bring a vehicle up to Loch Eriboll, unload some of the scrap and return with a replacement engine. After the decision was made, he was happy; he and Tony could fit the new engine, Simon and I were told we could leave if we wanted to. This was not the way I had envisaged the end of our Fair Isle adventure, although it had been profitable in more ways than one. I had found a partner I liked and trusted, while gaining invaluable expertise.

4

Fair Isle Versus the Rest of the World

'Well, that's it. We're on our own now,' said Simon, when we were back in my cottage in Fife. It had been perfect timing. John, who I knew from my days in the Firth of Forth, had been staying in the cottage and had recently left to work abroad.

Simon was reading over the letter we'd just received from Tom Henderson, the curator of the Shetland Museum. He had met him when he worked on the Armada wreck on Fair Isle. It gave us the encouragement we needed.

I am very interested to hear of your project for next year at Fair Isle . . . but somewhere round Fair Isle I am pretty sure that some homeward-bound Indiaman was lost. Haa James, who was dead before you ever came to Fair Isle, had the remnants of a most intriguing old tale of an occasion when the grass above Lestit was spread with bales of silk, unrolled to dry. So go and have a hunt and good luck to you! I am sure you will treat anything which you may find of archaeological or historical interest with due respect. Please keep in touch and if there is any way in which I can help, needless to say I will.

The Indiaman referred to was a Dutch East India Company ship of the seventeenth or eighteenth century that traded with the Far East and was often loaded with goods of value and interest. The *Adelaar* on Barra had been an outward-bound Indiaman, but this one would have been inward-bound, if she was carrying bales of silk. I hoped we might find other exotic cargo from the Dutch East Indies too, along with her own bronze cannon.

'It couldn't have worked out much better,' I replied, smiling at Simon. We had begun to set in motion our plans to dive for scallops over the winter, hoping to earn enough money so we could spend a summer on Fair Isle.

'What've we got in the way of diving gear?' he asked, as if reading my thoughts.

'I still have most of my stuff. An inflatable boat and outboard is all we need – maybe a decent vehicle.'

'Let's get a new inflatable. That old one we used with Chris was a nightmare,' Simon suggested.

I had been diving for less than two years and felt elated by the thought of working a whole summer in the clear water off Fair Isle, with the chance of finding a good wreck. We'd been given a pamphlet by one of the islanders that showed eighty-five known ships that had been wrecked on or near the island.

Simon also wrote to the Salvage Association and various insurance companies that we thought had insured some of the wrecks. One, the steamship *Duncan*, had been wrecked in 1877, the Union Iron Steamship Mutual Insurance Association of Dundee paying out on the loss. Unfortunately, this business had been wound up in 1891, leading to another dead end in our quest to purchase it. The *Canadia*, the obvious choice to buy, had a Danish owner, but we got nowhere when we wrote to them. As she had been commandeered before her loss, we approached the War Risk Insurance Office, but with no success.

'What did Chris do with the scrap you recovered from the *Canadia* when you were on Fair Isle?' I asked Simon, suspecting Chris had tried to buy it but, like us, failed to find an owner.

'He had a trailer full of scrap and we took it up to Lerwick to visit the Receiver of Wreck. We went to his office late on a Friday afternoon and filled in all the forms before he asked where we had it stored. Chris said he thought the Receiver had to store it in a safe place and we would carry it up the stairs to his office.'

'What happened?'

'The guy knew we were being silly. He smiled, tore up the forms and said f**k off and that was it. If we'd had a lorry load, it would have been different The value would have made it worthwhile for the Receiver to rent a storage shed.'

'Let's send a letter to the Receiver of Wreck in Shetland with copies of all our research and attempts to contact owners.'

This resulted in a non-committal reply, but a phone call was more positive; he had no issues with us working the wrecks if we kept him informed.

When it came to equipment we bought a second-hand Halflinger; it had a winch on the front for pulling a boat up the beach, or for hauling itself out if it were stuck. Much smaller than a Land Rover, almost the size of a Mini, it was ideal for the island ferries. The four-wheel drive would take us down remote tracks to beaches and at £200 it seemed a bargain. We'd already bought a new thirteen-foot six-inch inflatable boat, which would fit on top of the canvas roof of the Halflinger. It exceeded the width and length of the vehicle but was within the law – possibly. The diving cylinders, compressor and equipment were stowed in the back.

The heavily laden but purposeful vehicle looked the part as we set off for Oban to board the ferry. Arriving on Barra we moved into digs near Northbay. The cost of living in a B&B with all meals provided was minimal considering our earnings, and the regular meals improved our work output. The importance of good meals was something we remembered when we struck it lucky with salvage; Simon would always try to make arrangements to make sure we were well fed, our output as well as our enjoyment being increased.

After we had been working for a couple of weeks Chris and Tony came to see us, having returned to the island with the re-engined *Dewy Rose*.

'What about going out to the *Politician*?' asked Chris.

Simon and I looked at each other. 'When?'

'How about Sunday? It's just a fun day and we'll see if the islanders have left us any whisky.'

The *Politician* was referred to as the *Cabinet Minister* in Compton Mackenzie's book *Whisky Galore!* In February 1941 she left the Mersey for a voyage across the submarine-infested Atlantic, bound for Kingston, Jamaica, and then New Orleans. Two days later, in a south-westerly gale, she struck the rocks off the island of Eriskay, which lies north-east of Barra. The cargo of interest was 22,000 cases of whisky, worth around £500,000 at the time of loss. It was for export and no excise duty had been paid on them. The whisky was stowed in Number 5 hold but as soon as the ship was wrecked oil from the fuel tanks created a thick oily layer on the seawater within the hold. A salvage vessel attempted to save the ship by first discharging some of the cargo and then trying to refloat her. After most of the good cargo had been recovered – except for the whisky that lay untouched below the oily water – the salvage vessel sailed for port to give a report. Meanwhile customs officers sealed Number 5 hatch in an attempt to deter the locals.

The islanders had for centuries recovered timber and other useful items that washed up on the beach. They considered saving or rescuing these otherwise abandoned articles as a natural action and the whisky was no different. They set to work on the ship, pulling up thousands of bottles through the oily water, to be consumed or hidden, while customs officers attempted to stop them. Customs recovered little of the whisky, but a number of islanders were charged with stealing and a few sent to prison. Determined not to be beaten, Customs resorted to employing the salvage company to place explosives in the hold.

Again, a local fisherman told us where to search for the wreck and it was easily seen from the surface through the beautifully clear water, although with the kelp attached it gave the appearance of rocks. As the *Dewy Rose* lay above us, Chris and I searched for brass, while Simon and Tony headed for areas of the hold where any whisky might remain. After the loose scrap was recovered, I swam over to join them. The oil from her original loss had long gone and now, with near-perfect visibility, we could see the hold was littered with broken glass mixed with the old-fashioned telephones that were part of the cargo. Raking in the corners we found some complete bottles – most were empty, but a few had corks still in place and contained whisky. By the end of the afternoon we had recovered five bottles each, though some had a wisp of a black cloud beneath the cork, which we assumed to be seawater that had entered the thirty-year-old bottles, or a taint caused by the cork rotting.

That evening we were invited up to a remote croft on the west side of Barra by Peggy Angus, an artist Simon knew. We took a good carry-out of beer, along with a whisky bottle to show our host. As the evening wore on, the bottle from the *Politician* was brought to the boil, cooled and consumed. For me it was 'cheers' before holding my breath and hoping that it would not destroy my taste buds forever. Worse medicine would be hard to find and the suffering the next morning was unlikely to be outdone.

Profitable scallop beds were getting harder to find on Barra; the small scallops that we had left on the best beds would need several years to grow to a good commercial size. There were plenty scallops scattered over the seabed but they had to be sufficiently close-packed and in water shallower than 60 feet for us to pick them up quickly enough to make good money. By November, after we dived on the *Politician*, we knew that we would have to move north from Barra to the Uists in the hope that we could find better grounds. Finishing on Barra, we returned to our parents' homes for Christmas.

In January we set off in the ferry for Loch Boisdale in South Uist, finding good beds as we continued to move north through the Hebrides until we had earned enough money for our summer on Fair Isle.

When we returned to Fife in March my parents had a pile of letters waiting for me. Although I had not seen Peter I knew he was now working in the oil industry, diving in 500 feet of water off West Africa, his letters continually asking, 'Are you sure you don't want to go offshore? It's easy money!' But I always said no.

It was less than a year since we had worked together on Barra but rumours had percolated back from Fife – his future father-in-law, concerned about his daughter's relationship and Peter's habit of leaving his battered van on his drive, was heard to say, 'When's Neptune going to remove his old van from my drive?' I was not sure the *in-law* relationship was going as well as he had hoped but there was no hint of that in any of his letters. He did mention leaving Dan, getting engaged and diving for anemones for a living before diving in the oil industry.

I also received a request in his letters:

> One thing, Alec . . . could you do me one favour? Communications, particularly mail, are really atrocious; my girlfriend sent me a parcel a few weeks back which simply hasn't arrived, for instance. Do you think you could arrange for some red roses (now isn't that romantic!) to be delivered to her for Christmas? I daren't risk trying to

send anything from here in parcel form. I expect there's an Interflora shop in Cupar that would deliver them, and if you could send me the amount I'll post it on to you. If they're not too much, two dozen red roses would be great, and it would be nice to know that she was getting something from me for Christmas. If you could have them print something like, 'With all my love, Peter' on the card, that would be fine.

I went off to buy the roses with much teasing from Simon. 'You'll never get paid, Alec,' he declared.

'But he's relying on me to do it,' I groaned.

'I'll put a bet on that he'll not pay you,' said Simon, laughing.

I knew Peter was easy come, easy go. If you wanted to borrow £50 and he had it he would lend it to you and forget about it. If it was the other way round, well . . .

'I'll not take a bet, but I'll do it this time,' I said. I knew in my heart I would not get paid and that I would do it again if he asked. Two dozen red roses were duly ordered, paid for and sent to his girlfriend.

Simon was right: I never did get paid, but Peter made up for it in bundles much later on when we damaged some of our equipment and he was immediately able to ship us replacements.

The boat trip to Fair Isle was always bad. Strong tides run between Shetland and the island, and add to that the Atlantic swell and it only requires the gentlest wind blowing into the current to create very steep waves. This day was no exception.

I was glad to arrive on dry land after the two-and-a-half-hour trip. We helped unload our equipment onto the island lorry: our kitbags and clothing went to the hut that was to be our home, and our diving equipment was placed at the side of the road at a point nearest to the first wreck we would work.

Simon had arranged our accommodation with one of the islanders. It was a simple wooden hut made from a disused part of the old bird observatory, which had been rebuilt on a croft near the post office. Creosoted many years before, the preservative gave it a dark grey colour. It had one door and a single window providing light. Running over the roof was a steel cable anchored deep in the ground to prevent the hut blowing away in a gale. It contained a sink with running water and electricity connected to the Fair Isle system that ran from six in the evening until ten at night and for a short time in the morning. There was no toilet, but plenty of countryside.

The hut gave us peace and privacy. Life was simple, to be lived as the weather dictated. The excitement of our arrival and the thought of diving on the wrecks in the beautiful clear waters spurred us on to organise the equipment in anticipation of working as soon as possible. Having unpacked our bags, we walked across the fields to look over the cliff at Malcolm's Head. From there we could see the wreck site of the *Canadia*. The sea looked fearful, a heavy westerly swell pounding in over the rocks, among which hundreds of seabirds circled as they flew to and fro beside the ledges on the cliffs. Tired from the long day, we moved back from the edge and, sitting on some stones, made our plans. This was the transition point from dreams to reality: it was up to us. The more time we spent in the water looking or working wrecks, the better our luck would be.

Wandering south along the top of the clifftop, we reached Hestigeo, where we had kept and launched the inflatable the previous summer. Finding a good place to store our diving gear, we went to the nearest croft, which was called Lough (pronounced 'low'). Jerry farmed the croft, so he was known as 'Lough Jerry'. The people of Fair Isle were always welcoming and he was no exception. He invited us in. Sitting in his clean and tidy croft with his wife Aggie, it became obvious that if Jerry had been forty years younger he would have been diving on the wreck with us, particularly if he saw a few pounds in it.

After Jerry had given us information on the tides and swell, to determine where and when it would be most suitable to work around the island coast, we finally asked his permission to carry our boat and equipment across his land to Hestigeo and leave it at the corner of his field.

Our life slipped into a comfortable routine broken only by bad weather, or at times not-so-bad weather when we just wanted some time off. We were both aware that we must keep some pressure on ourselves, to stop us treating these months as a holiday, but it became easy to miss a day if any form of interesting activity came along, such as sheep shearing, or an offer to visit the lighthouse, or go fishing.

'What do you think, Sy?' I would ask.

'It's not great weather for diving,' he might reply.

'Let's have an easy day, then,' I would venture, having gauged his feelings. He at times would ask the same question of me.

If we had been busy and needed a break, we liked to explore the island; it was beautiful, it was romantic, and it was not difficult to fall in love with it. In windy weather the simple pleasure of spending an hour lying at the top of a cliff, watching the birds and dreaming, was hard to beat.

The *Canadia* became a test ground for improving my ability with explosives. I now used a specialist submarine blasting gelatine made by the Nobel Explosive Company, as it was called at the time – it was founded in 1870 by the famous philanthropist Alfred Nobel! Ironically, Nobel (of Peace Prize fame) made much of his money from selling munitions. This explosive was far superior to the 'quarry' explosives and I was able, by accurate placing, to reduce most charges to just ounces of explosives following Chris's methods.

With diving completed for the morning, our loaded inflatable was driven back to the small pier at Hestigeo. There we threw the scrap brass from the boat onto the pier: twisted portholes, copper pipes, damaged sea valves and numerous small pieces whose original purpose we had no idea. We carried the scrap up the steps to a growing pile, where we estimated the weight and noted the figure. We might then wonder about spending the afternoon searching for wrecks. If the weather was good, we could go anywhere around the island.

'There's the *Black Watch*,' Simon might suggest. Lost in 1877, she was a sailing ship of about 1,300 tons. 'Or how about the *Lessing*? We might find some of its cargo of china. We can motor round the south end of the island with the inflatable to get to the site.'

We had a photograph of the *Lessing*, showing her up against the cliffs where the Fair Islanders had managed to rescue the German crew, along with 465 German immigrants, when she had run up against the island in 1868 in a thick fog.

'We could have a look for the *Duncan* on the way,' I might offer in reply. 'The sea is rarely so calm at the south end.' I liked the *Duncan*. She was a steam wreck, and there were bits of brass lying about with little left of the hull, which had been destroyed by the exposed position in front of the lighthouse.

'OK. A quick look, and if we don't find much more we'll carry on round to the *Lessing*.'

It was always pleasant to lie in the sun waiting for the air bottles to fill. The days were long at the height of summer. There was no hurry and this break gave us the required time above water to avoid any decompression problems. We always tried to make the deep dives first, following with shallower dives to act as decompression. If either of us had a bad bend on the island it would be serious, possibly fatal, because there was no decompression chamber available in Shetland.

Finding wrecks was made easier by the islanders, who would point us towards known positions and areas. The seabed was as varied as the land, its surface covered with thick growths of kelp and oarweed that moved backwards and forwards in the swell as if blown by the wind. This made searching difficult, as it obscured the bottom. The diving was still fun: there was always the unknown as we searched for older wrecks, hoping we might find something of value or interest. Sometimes when we knew where a wreck had been lost we found nothing and assumed it had been ground up by moving boulders in the winter gales. On a rock seabed containing deep wells we found the propeller and rudder of the *Duncan*, the largest remaining parts of the ship.

At the site of the steam trawler *Strathbeg*, Simon found its bell sitting on top of the engines. It is now in the Fair Isle museum. Otherwise there were few souvenir-type finds due to the exposed sites. On trips around the island we would regularly see dolphins and seals. Once a school of orcas came close into a wreck site; it was impressive to see, and I was fortunate it was Simon's turn to dive.

When our day was finished, we wandered up to our hut to get a meal. If we were lucky, Deidre, who had become a regular visitor, cooked us a good dinner. She was there for the summer, paying her way by cooking at the observatory while taking advantage of the island and its bird population. We enjoyed her company, as well as her food. It reminded me of the B&B in Barra and how much easier and more enjoyable life was when we came back to a warm home, a cooked meal and a fresh face.

Simon's intention to go back to a steady job after this final fling was often on my mind. The more I worked on the wrecks in these beautiful waters, the more I enjoyed it and wanted to continue. I was undoubtedly influenced by the idyllic setting and the total lack of pressure. The island was always busy and the mailboat, the *Good Shepherd 2*, went to Shetland twice a week to fetch supplies. When the boat returned it was a social occasion, with most of the able-bodied islanders gathering at the pier to help unload. Unspoken, but perhaps implicit on the island, there was a duty to contribute as an entity. Fair Isle was a busy, happy place, and it was a pleasure to be part of it. There were plenty of children at school, a new mailboat was arriving and the community pulled together for the benefit of everyone. It was as if each individual had a special skill, not only for raising their own family but for contributing uniquely to the survival of island life as a whole. One weekly event we never missed was the football match on Saturdays, Fair Isle versus the Rest of the World. Simon would have stopped lifting gold bars to attend.

We were often invited to people's homes in the evening – something we both looked forward to and enjoyed, along with the bonus of the opportunity for a warm bath. Conversation was always easy; we would enquire about the latest island news and would, in turn, be questioned about what we had found and recovered. The scrap lying outside our shed could be seen by all, its origin and sometimes its purpose discussed.

One evening during a visit to the home of Stewart and Triona, two islanders we knew well, the island of Foula was mentioned. It is Britain's most inaccessible inhabited island. Lying west of Shetland, all I knew about it was that it had a poor ferry service and a rough gravel airstrip that was seldom used, except in emergencies. Stewart and Triona showed us a rare book by a Professor Holbourn about the island. The writing was judged to have romanticised the island considerably, but it included a section on the wreck of the largest passenger liner in the world, built in 1899 – the *Oceanic*. I could hardly believe that there was such a large wreck lying so close and I had not known about it.

The *Oceanic* had been wrecked two and a half miles off Foula on a submerged reef and an attempt was made to salvage her. The book states:

> The launch of the *Lion*, a salvage boat which hurried to the scene, was capable of a speed of 10 knots, yet she was unable to make any headway against the tide, although she tried with full steam for fifteen minutes. Even then it was not the top of the tide, and the officer in charge reckoned that the full tide would be 12 knots. He confessed he would not have believed it if he had been told. Before long every available islander was commandeered to assist in the work of salvage. Troops and guns were safely removed and an attempt was made to take off some of the more valuable furniture and books. The salvage boats and the trawlers made 'a bonny sight' with their searchlights that lit up the hills. But before long a gale blew up and the great ship slipped below the waves. We often think longingly of the many tons of coal, blankets, linen, pianos, and all the furniture of the luxury liner lying just beyond our reach, so near and yet so far. Although the *Oceanic* has altered her position and now lies in deeper water she can still be seen beneath the waves.

My eyes lit up at the thought of the wealth of copper and brass to be found. There was a caveat: the book mentioned that the tide ran at up to twelve knots over the site – that would make it almost impossible to dive, let alone carry out any salvage work. I ignored it as an exaggeration. I was in no mood to have my enthusiasm dampened.

'How about it, Sy?' I asked, full of new-found hope.

'It'll be a major problem to get there,' he replied.

'If we could just get a look at the wreck and see if it's workable . . .' I suggested. 'It surely wouldn't cost much.' My blood was stirred. I was committed, but it was easy for me – this was exactly what I had been looking for, while Simon had been planning to return to a career in journalism.

Luck was on my side. Deidre told me that the bird observatory had chartered a Loganair aircraft to take ornithologists into Foula. The Islander, carrying a single pilot and nine passengers, would take a party there in the morning and return in the afternoon to pick them up. There remained one spare seat, which I quickly booked (thus reducing everyone's fare, as the cost was equally divided among the passengers).

I was brimming over with excitement at the chance of going to Foula and the possibility of working the wreck of the *Oceanic*. I could think of nothing else. Knowing Simon was lukewarm about it, I attempted to conceal my enthusiasm in case I put him off, but if it looked like a viable project, I decided to reopen the subject.

The flight required good conditions to land on the rough gravel airstrip. Getting to Foula was known to be difficult – or, more accurately, almost impossible within a reasonable time. The Foula mailboat was weather-dependent and if I had caught it I would have had to stay on the island for at least a week, requiring a tent, as I was told there was nowhere to stay except one old croft house that was normally let for long periods to birdwatchers. On the day I was up early with my coat on and a notebook in my pocket, then Stewart came from the post office to tell me the flight had been cancelled due to bad weather. My heart sank.

We used the time to continue researching the *Oceanic*. Simon enjoyed this aspect and was warming to the project. He gleaned some information by sweet-talking his old newspaper contacts and learned she had been a beautifully elegant ship with a long, graceful hull, slightly raked bow, a counter stern, with a low superstructure and two tall, well-proportioned funnels. Popular on the transatlantic route to New York, she was transferred from her home port of Liverpool to sail from Southampton in 1907. Operated by the White Star Express Service, she shared the route with the slightly slower White Star ships *Teutonic*, *Majestic* and *Adriatic*. The *Olympic*, *Britannic* and *Titanic* were built to replace those three ships, but the *Oceanic* was to be retained as a relief ship. At maximum speed she would burn over 700 tons of coal a day. On her maiden voyage, she left Liverpool with nearly 2,000 people on board, bound for New York. The 2,780-mile trip took six days, travelling at an average speed of 20 knots, and she arrived to a rapturous reception. Tugboat owners vied for the honour of pulling her into position. She soon became known for her reliability in arriving on time at her destination ports.

The *Marine Engineer* magazine's description of her was full of praise:

> The saloon has as many seats as the ship has first-class berths, so everyone can dine at the same time. The room is remarkable for its decoration. It is panelled in oak washed with gold, while its saloon dome (designed by a Royal Academician), decorated with allegorical figures, representing Great Britain and the United States and Liverpool and New York, is very beautiful and striking.
>
> The smoking room is lighted by two domes, one on each side of the middle line of the ship. Its ornamentation is tasteful. The pictures depict various scenes in the life of Columbus, and in the corners are niches containing beautiful Italian figures. The walls are covered with highly embossed leather gilt, and there is a carved mahogany frieze. The ornamentation, of both of the frieze and of the stained-glass sliding shutters on the side ports, is of sea nymphs.

This description is backed up by Commander Lightoller, who was the most senior officer to survive the loss of the *Titanic* and an employee of the White Star Line. He recalled the *Oceanic* as the finest ship he ever sailed in. In his book *Titanic and Other Ships*, he wrote:

> I got my severest mailboat training during the seven hard though happy years I spent in the *Queen of the Seas* [as the *Oceanic* was then called]; a wonderful ship, built in a class of her own and by herself.
>
> The usual custom is to build twin ships, as with the *Britannic* and *Germanic*, *Teutonic* and *Majestic*. Then in lone and stately majesty came the *Oceanic*. She was an experiment, and a wonderfully successful one. Built by Harland and Wolff, regardless of cost, elaborate to a degree, money lavished where it was necessary, but never gaudily as is so common nowadays. Her smoke-room doors were a masterpiece in themselves and cost £500. There was eighteen-carat gold plating on the electric light fittings throughout the saloon and staircase, and paintings by well-known artists worth a cool thousand apiece; hand carvings of delicate work and the joy of souvenir hunters. Every deck plank was pickled wood.

The *Oceanic* had been designed and built with reinforced decks to take 4.7-inch guns, which were fitted at the start of the First World War, along with loading 7,000 tons of coal and a complement of 600 to crew the ship. She was then sent to patrol the northern passages around Shetland. With a length of 704 feet and breadth of 68.3 feet, she would be a massive wreck. *How could we fail to find it?*

When the weather improved, the bird observatory rearranged the flight. Some of the original passengers had left the island, but newcomers had taken their places. Standing on the landing strip on Fair Isle, I heard the aircraft before I saw it appear from around the side of the hill making a neat landing prior to turning and disembarking a batch of birdwatchers from Sumburgh. I took a seat at the back, being the odd person in a group of ornithologists. The two engines were started and the aircraft taxied to the end of the runway, where the pilot gave it full throttle. We bumped along before soaring up over the sea. I felt the flush of adrenaline hailing the start of a new adventure. This was our great chance. My plan was to obtain information on the position of the wreck, as well as arrange a place to stay. I dreamed of things to come, my preconceptions of Foula and the *Oceanic* creating images of paradise and wealth.

Within minutes we could see the outline of an impressive island ahead of us. It was slightly larger than Fair Isle, and the west side had steep cliffs dropping vertically into the sea; the east, with smaller cliffs, had a small inlet defining the coast. Everyone's eyes were trained ahead, mine veering off to a place two and a half miles east of it, where I could see the crests of the waves disturbed by the sea. That must be the submerged reef where the *Oceanic* lay. It looked okay for diving; the island gave it some shelter from the west and north-west. The sea looked calm, but it always looked quiet from the air; it was only when you were down at its level, looking straight at the waves, that you could appreciate their height. My eyes moved from the sea to the island, quickly growing in size as we approached.

The gravel landing strip started at the edge of a cliff, and the aircraft lurched in the turbulent air as we passed over it. Within seconds the wheels were on the ground, my seatbelt pulling tight as the brakes were applied. There were no buildings or anyone to meet us. We disembarked from the plane and watched the pilot taxi back over a rise towards the cliff edge, where he disappeared from sight. Next thing he was tearing past us again on take-off.

We stood and looked at each other as if we were on a deserted island, the silence only broken by the call of the kittiwakes.

This was Foula. I liked it.

First, I had to find the South Biggins croft, as I had been told on Fair Isle that the family living there might help me with my quest. It was a quarter-mile walk from the airstrip and I assumed it would be easy to distinguish their croft from the other five in Da Hametoon as the door and windows were painted gaily in light colours.

Entering the boundary wall of Da Hametoon, I noted some crofts that looked as though they had been abandoned many years before. They had an air of neglect about them. The turf or thatched roofs had fallen in and were now inhabited by fulmars. The birds made aggressive noises to warn me off. A roofless church stood in its graveyard, but near it was the brightly coloured and tidy South Biggins, where smoke rose from the chimney and the smell of burning peat conjured up memories of the Western Isles, with their warm and friendly hearths.

I knocked on the door with a slight feeling of trepidation. I hoped somebody was in. So far I had not seen a soul – though I had been made aware that the islanders were afraid of catching colds from *outsiders*. Apparently they had little resistance and were susceptible to both catching the virus and being severely affected by it.

I heard voices on the other side of the door before it was opened by a man who looked to be in his early seventies. I guessed this must be Bobby Isbister, the person I was looking for.

After introducing myself I explained my interest in the *Oceanic*, mentioning I had been working on Fair Isle. He beckoned me in through the small porch into a neatly kept room where I was introduced to his wife, Aggie-Jean, and his son, Eric, who was in his thirties. A good-looking young man, he was slightly nervous at my presence. I sat on the left of the stove and looked around – the walls and ceiling had a dark surface, caused by the peat fire and the Tilley lamp. It reminded me of a room at home, where the coal fire and my father's smoking had caused similar discolouring. The low-set peat stove stood on curved Victorian-style legs; the kettle sat on the oven side to keep it warm and Aggie-Jean pushed it above the fire to make it boil. Bobby and Aggie-Jean sat opposite me on the resting chair, a long bench-like piece of furniture that could be used as a bed. Eric sat on a single chair on my side of the stove. I repeated my reason for being on Foula while Aggie-Jean fiddled around with cups and placed some scones in the oven.

The island was too remote to have electrical power cables laid from the mainland, and I had not heard any generators running. I did notice on the way to the house that there was a well at the bottom of the slope and wondered if each house had its own well for its water supply.

Bobby, sounding like an experienced storyteller, explained how the islanders had helped to unload stores from the *Oceanic* when she was stranded on the reef called the Shaalds (shallows).

'We've an old teapot and tablecloth embossed with the White Star emblem recovered by my father. I can't put my hands on it but it's somewhere in the house,' he said.

He not only knew about the wreck but he was interested in it.

'We'd have taken more useful bits,' he continued, 'but the tobacco bond was found and everyone filled their pockets. When the weather got worse, the ship just disappeared one night after two weeks on the reef.'

The reef was always submerged, with two distinct rock heads that were the shallowest parts before the seabed sloped steeply into deep water. The search area would not be too extensive, even if the ship had slipped into deep water: as Holbourn's book had stated, she was such a big ship at 17,000 tons that some parts must be shallow enough to dive on. Most of the wrecks we worked had broken up where they lay; perhaps this was the case with the *Oceanic*. A ship took its shape from thousands of steel plates riveted or welded together; it collapsed in the same way as a house of cards, the plates tending to flatten out along the seabed, thus losing any form or height. Over time they gradually corroded, leaving the engines, stern and bow section as the most recognisable parts.

'D' you think we could find the wreck?' I asked.

'The tide'll stop you,' said Bobby. 'It's too strong. Faster than our small boats.'

'There must be a slack time?'

'It's a bad area, catching the swell from the south-west. I'm not thinking there will be much peace for you on the Shaalds,' Eric said, shaking his head. Both Aggie-Jean and Bobby shook their heads in agreement.

'We could surely get a look at it,' I persisted. 'Even if it's just for a few minutes?'

'It's a strong tide,' repeated Bobby, politely trying to discourage me.

They appeared genuinely concerned that I should not put myself in danger. I reasoned there must be a time when the tide turned and the current would fall away before changing direction. If we used that period of slack water, it was worth a try. My mind was made up. In fact, it had been made up before I met them. I wondered what they thought of me, a complete stranger, coming into their world with the intention of working a wreck that had remained untouched since she sank in 1914 and now, in 1973, nearly sixty years later, I was interested in trying to find it.

The Isbisters explained some of the island landmarks and gave me directions to Ham, where I might get help with accommodation. Following their instructions, once past the manse I found a small quarry which supplied the soft stone for the road. Taking the right fork led me directly to Ham Voe (Ham 'creek' or 'inlet'). Once there I followed a steep path running down to a burn. Looking seaward, I could see the burn flowing into the voe, the pier almost cutting the inlet in half before it widened into the sea on the east side of the island. I crossed the wooden footbridge over the burn, climbed up the path on the other side, past Ham croft, before rejoining the gravel road that led to the pier.

Approaching the pier I could see the island mailboat, an old RNLI lifeboat, sitting on a cradle at the bottom of a slipway. There was a substantial crane for lifting it out of the water onto the cradle which ran on rails, allowing it to be hauled up well clear of the sea in bad weather. Two small open fishing boats lay on moorings, sitting within the pier's shelter. The place looked perfect for us. Everything seemed to be falling into place. I sat on the parapet of the pier, drumming my heels against the stone wall like a small child waiting for something to happen. My thoughts went beyond working an inflatable boat, as this project would require a salvage vessel. Decisions would have to be made: whether the boat could be based on Foula or have to be operated from the Shetland mainland.

I sensed I was being watched and turned to see an old man with a pipe grasped firmly between his teeth standing on the road. He stayed there, gently nodding his head. I waved in acknowledgement and walked up the slipway towards the nearest house. The house looked empty: no smoke came from the chimneys, the path to the door was unused, and sheep grazed up to its walls. I was reluctant to look in the windows, just in case there was someone living there. Unlike most of the island houses, it was tall with a full-height upper floor, small windows and a slate roof, whereas the other houses I had seen were one and a half storeys and roofed with bituminous felt. It looked to be watertight and the number of windows suggested it had five rooms: three upstairs and two downstairs. The gable ends had two chimney pots each: at that moment herring gulls sat on them, leaving white marks of droppings trailing down the slates. Shallow pools of water lay on the ground next to the walls, as if drains were blocked. On the eastern seaward face, a square tower had been built as an extension; it looked Gothic and unfinished. I knew immediately from Professor Holbourn's book that this must be the house in which the 'laird' stayed when he came to the island. '*The laird started a grand new porch to the Ha, but alas the war came before it was finished.*' That had been the First World War and the porch still looked unfinished.

Behind the house were the remains of a walled yard with a small stone shed that possibly contained peats, as there was no outdoor peat stack. A badly corroded galvanised water tank sat on a frame attached to the north gable of the house; this appeared to be the water supply. I examined it closely, noticing drips coming from rusty blisters in its bottom; the water fed into it from the slate roof, where it was collected by the rhones. There were no electric cables to indicate an electricity supply. I liked the house. Being next to the pier, it was perfectly placed. I decided I would ask about it; the Isbisters had told me that Elizabeth, at Ham croft, had a small cottage she rented to birdwatchers and she would surely know about this house.

Retracing my steps for the few hundred yards to her croft, I knocked on the porch door. The knock caused the uncontrolled barking of a dog, followed by the sound of a woman scolding it as she approached the door. I prepared myself for the dog, hands ready to fend it off as the door opened sufficiently for a head to look out.

She looked me up and down before saying, 'Come in, come in, the dog'll not harm thee.'

The door opened fully and the dog shot out, barking and prancing excitedly, before following me quietly into the house. On the resting chair next to the fire sat a frail, elderly lady with long, delicate, grey hair, who looked well into her seventies. She gave a slight smile and introduced herself as Joann, giving me an encouraging nod.

'I'm Alec,' I said, as I patted the overfriendly dog. 'I've come from Fair Isle, where I've been diving on wrecks.'

Elizabeth cocked her head to one side as she listened. She had a squint that was disconcerting until she smiled. It was a large smile, a 'pleased-to-see-you' smile.

'Tea or coffee?' she asked. It was said in the way that was an order rather than a question. I had been well fed at the South Biggins, but I daren't have said no.

'Yes, please. Coffee, with just milk,' I replied.

'You'll have some of my soda bread and butter, too?' she asked, as she buttered the bread. She must have been in her fifties: a short, stout woman, wearing thick stockings, a brightly coloured cotton skirt and an apron, giving me the impression of a motherly person, although there was no sign of children. We had the usual conversation about health and the weather before I explained about my quest to look for the *Oceanic*.

Elizabeth preferred to ask a stream of questions – where I was from? Where had we worked? What was it like working on Fair Isle? I answered politely as I looked around, seeing dried fish and salted mutton hanging on a rail over the stove. Behind me, steep stairs led up to what I supposed must be loft-type rooms similar to those I had stayed in on Barra. The house looked too small for paying guests.

'I'm told you might be able to help us with somewhere to stay,' I suggested as I finished the coffee and buttered bread.

'Well, well, what can I do?' she said, looking towards Joann.

I waited for a reply. I was not going to hurry her. 'We're not fussy,' I said. 'Almost anything will do.'

'I might be able to get the Haa,' she said.

'Where's that?' I asked.

'It's set above the pier,' she replied. 'It's the only empty house.'

'That'd be perfect,' I said, trying to supress my excitement. It *was* the house I had looked at. My heart beat faster as I willed it to be available.

'Mind, I can't promise,' she said. 'I'll have to speak to Phil. He's a member of the family that own the island.'

She explained that the island had been left to three brothers: one had died and the other two still came to the island occasionally. Alasdair had a house at the north and Phil used the Haa, the traditional laird's house. They would not be coming to Foula this late in the year, so if we could carry out some repairs they might be prepared to let us rent it for a few weeks. 'Mind, the house has no electricity or water supply.'

The idea of repairs would suit us, as we could do them ourselves at little cost. Any rent was likely to be small and would not knock a hole in our finances. I did not know if we would find the wreck, let alone work it, but a few weeks would be more than enough to get us started. If it worked out, we could consider a longer-term arrangement.

'Call me after a few days . . . maybe a week? I'll let you know what he says,' Elizabeth promised. She was peering at me, as if she wanted to be convinced I was a responsible person. Meanwhile, Joann looked pleased.

I made my move to leave, but Elizabeth would not have it and encouraged me to stay for a while longer. 'Another cup?' she asked. Our conversation continued like a gentle sparring match as we both tried to find something out about the other. I thought of Simon: he was better equipped for this than me. Old Joann sat by the Rayburn knitting a shawl, occasionally nodding in agreement; so far she had said nothing, but I thought she might remember the *Oceanic* if she had been on the island at the time. It was not until much later that I learned she had been married to a diver who'd made a failed attempt to salvage the wreck in the 1920s.

Finally making my move to leave, I thanked Elizabeth and

Joann for their hospitality. I intended to go down to the Haa now, armed with the knowledge that it was empty. I could peek through the windows at least and see what sort of state it was in.

Gentle sparring aside, I felt Elizabeth was sad to see me go. She came out of the door along with her dog, Dusty. 'You've got my number?' she repeated several times to make sure I hadn't lost it. 'Mind and call me.'

'I'll call you, don't you worry about that,' I said, before wandering down to the Haa.

I pushed my face against the window: the inside of the kitchen was austere and looked cold with its concrete floor and paint flaking off the stone walls. The other room, a sitting room, was dark, with old furniture and stained wood-panelling going halfway up the walls. After another stroll round the outside of the house, I went down to the rocky shore, just a stone's throw from the front of the Haa, and sat on the rocks. Man-made pools had been cut into the rock. I wondered what they had been used for – perhaps to extract salt or wash fish? I wasn't sure. Looking out to sea, there was the broken area of white water caused by the tide running over the submerged reef that the *Oceanic* had struck, the Shaalds of Foula. I sat for a long time, my eyes on the turbulent water, trying to imagine the *Oceanic*, that vast ship, stranded on the reef.

In contrast to Fair Isle, Foula appeared uncared-for. Derelict crofts were everywhere. Many people must have left and now the island's standard of living looked extremely basic. I knew little about the island and wanted to know more; on the mainland it was spoken of disparagingly, which did not chime with my experience of speaking to Elizabeth or the Isbisters. I was puzzled by it, but I had only seen a handful of the twenty-eight people who lived there. It was a new world to me, with the land looking severe and unforgiving, for both livestock and people. Said to be the most inaccessibly inhabited island in the UK, perhaps its inaccessibility was part of the problem, but I liked the potential of it: the unknown. The harshness of the place that must have defeated those who had left, although I realised, of course, like Barra, some of them would have been pleased to leave, taking the opportunity to make a new life in another part of the world.

There was nothing else I needed to arrange, so I walked further up the coast, exploring another derelict croft with its drystone walls and collapsed roof. The thought of camping in the shelter of the walls while attempting to look for the wreck did not seem a realistic option; I knew Simon would not like the idea, but it could be a last resort.

I returned via the Haa to have a final look before choosing a slightly longer route to the airstrip past the school and post office, seeing an elderly woman feeding hens from an open sack of feed that she kept in a red telephone box. Next to the box stood a croft house with a small porch that was used as the post office. It looked neglected, with loose stonework stuck with bits of wool where sheep had rubbed against the walls. There was no sign of life within it, no postcards for sale, only a few faded post office notices propped inside the window.

After take-off the pilot flew up the centre of the island over the rock stacks at the north before turning west and flying past the second-highest vertical sea cliffs in Britain. They looked stunning, with a mass of birds on the ledges, some soaring in the updraft at the rock face, the heavy Atlantic breaking against their base, blurring the boundary between rock and sea. I thought again of our chances of success on the *Oceanic*. I was pleased that it had been wrecked on the east side.

Perhaps it was wishful thinking, but deep down I had convinced myself that Simon would go for it. I knew he had not applied for any jobs, whether that was him just being lax or an indication of a reluctance to leave the diving work, I wasn't sure – but I suspected it was the former. Either way it gave me encouragement.

A week later, we phoned Foula. Elizabeth had received permission for us to stay in the Haa, but the owner had suggesting we came in the second half of August. After I had asked about the availability of food and peats for the fire, I passed over the phone to Simon for him to exchange news with Elizabeth. We would ring her again, and after Simon's bit of Shetland gossip I was sure she would be looking forward to the call. With our recent research, Simon was now convinced we were looking for something exceptional and had become as excited as me.

We quickly arranged a charter of the *Good Shepherd* to transport our scrap to Wick, on the Scottish mainland. From there we could load it onto a lorry, transport it to Glasgow and sell it. The sale would realise sufficient funds to keep us going while we searched for the *Oceanic*.

5

Foula and the Largest Ship in the World

With our goods and chattels stowed on the *Good Shepherd* on its way to Foula and money in our pockets from the scrap sale, we tried not to give the impression of travellers, but that's exactly what we felt like. Also aboard were Deidre and a group from the International Voluntary Service (IVS). Our trip had provided a unique opportunity for them to visit the island. It was a six-hour trip, and within minutes of our arrival we were met by Elizabeth from Ham croft. Although she had confirmed on the phone that we could stay in the Haa, I was relieved when she reassured us that the house was available; it all seemed too good to be true.

As the tide was falling, the *Good Shepherd* might touch the rocky bottom beneath the pier at low water, damaging the hull, so the IVS visit and our unloading was limited to a couple of hours. We worked with the crew to get all our gear onto the pier. Tom, the elderly islander I recognised from my previous visit, stood on the pier smoking his pipe and watching us.

'Man, I'll tell thee . . .' he started, before explaining that the island had peculiarly turbulent wind conditions due to the high cliffs and hills on the west side. He claimed that you could light a pipe in one place and ten feet away the wind would pick you up or blow you over. Like listening to ghost stories, I was not quite sure whether to believe him, but taking his advice we tied the inflatable and other bits to metal rings on the pier. After the hard work of unloading was over, we were struck by the sudden quiet when the boat left and the engine noise gradually died away in the distance. We looked at each other, knowing we had committed ourselves to the project – there was no going back. We walked up to the Haa, where Elizabeth was waiting for us at the door beside our baggage.

'Well, there you are, boys,' she said, as she ushered us in. It seemed dark inside after the brightness outdoors. A damp smell pervaded the passage, as she directed us through to the living room. Passing the bottom of the wooden stairs, we turned right and entered a room that looked as though we had moved back into the past. The peat fire she had lit for us was throwing moving shadows behind the furniture. I was concerned that I might have over-egged the pudding when describing the Haa to Simon, but he looked unconcerned. The house gave us the opportunity to search for the wreck in reasonable comfort. I loved it: its age, its position, the ancient feel of it. I knew we were incredibly lucky to have such a convenient base on the island.

Closing the panelled door behind us, I instinctively reached for the light switch only to remember the house had no electricity. I looked around, letting my eyes acclimatise to the gloomy room, the dark varnish on the wood-panelled ceiling reflecting little light. I walked diagonally across to the other window. It faced the pier and was set into a four-foot-thick gable end; although larger in overall size, it had small panes of glass that restricted the light and the frame looked frail. I could imagine someone curled up on the window ledge reading a book on a wet, windy day. Next to the window was the open fire. The stonework at either side was partly collapsed, but even so the peat fire threw plenty warmth and light into the room. Above the fireplace was an oil painting. Moving towards it to see what it depicted, as it had been blackened by years of smoke from the fire, I asked Elizabeth.

'It's the north o' the island,' she answered.

I moved my attention to the remains of an old gas lamp fixed to the ceiling. The pipes felt loose to the touch, the mantles damaged and unusable.

'You'll need a Tilley,' she said. 'It should be in the kitchen. Mind, if it's not there I can lend you ma spare.'

'Thanks, but we've brought one,' replied Simon, as he felt the seat of the old mock-leather settee. 'The house looks ideal,' he continued, before asking, 'You mentioned a shop on the island?'

'Naw, it's nearly finished. You're better ordering from Walls [pronounced 'Waas'] on the Shetland mainland. They'll send a box with the mailboat.'

I looked down at the old woollen carpet that covered much of the floor. Holes, tears and damp patches were obvious, although partly covered by the mock-leather chair and matching settee. The wood beneath the carpet was visible around the extremities; it looked solid but had the occasional white mark of damp. In the centre of the room was a table with several wheel-back chairs. A bookcase and a chest of drawers added to the furniture. Above the dado rail that topped the woodwork, a canvas-type material bulged in places where the plaster behind it had crumbled. The material had been whitewashed but now it had a dull, almost yellow tinge. In the north-west corner of the room – the darkest – was a table with three large drawers. I drew back the centre drawer and saw a small cardboard box. I lifted it out and, after showing it to Elizabeth, opened it. It was full of small bits of string, slight twists on them as if they had been parts of knots. There was a note in the top. I took the note and walked to the window to read it. It said, 'Bits of string too small to be of any use'. I laughed – like the unfinished porch, the 'laird' must have saved his pennies.

'Are there many of the laird's bits and pieces in the house?' I asked.

'Na, small bits like that box. Anything else'll get ruined with the damp.' Elizabeth turned to go out. 'I'll leave you, but if you're needin' anything, come up to Ham.'

I followed Simon into the kitchen at the north end of the house. It had a concrete floor with a wooden table in the centre. An old gas stove stood near the sink under the kitchen's only window. I found the water pipe, which came through the wall from the corroded tank outside. That would be the first thing we would fix: we needed a water supply, even if it came from the roof. Going up the wooden stairs, Simon took the bedroom on the south side above the sitting room; I settled for the one at the other end of the house, whether it would be for weeks or years I could not guess; I looked out the small window towards the reef where the *Oceanic* lay. Between our bedrooms was a smaller room with two beds. The walls and ceilings of all the upstairs rooms were lined with bare unpainted wood and there were no signs of leaks from the roof. Looking round, I thought, *This is to be my home.*

I carried in all the gear from outside the back door and dug out the cast-iron kettle we had brought, while Simon went to Ham to fill a container of water. When he returned we sat in silence enjoying the heat and light from the open fire with the kettle perched on the peats beside the flames. I was not sure what to make of it all. Our move had happened so quickly and easily, it was difficult to believe we were on Foula. The weather was good and might give us a chance to find the wreck, although we had heard nothing but bad reports from Shetland fishermen about the conditions on the Shaalds, where the wreck lay.

'We'd better try and ask someone where the wreck is,' said Simon.

'The Isbisters are the best bet,' I said. 'They're the people I spoke to before. You might get more out of them.'

'Let's go before it's dark. I'd like to see more of the island.'

Closing the door of the Haa, we set off on foot to the south end. There were no cars on the island, but there were several tractors and we could not help noticing the remains of two scrap motorbikes lying randomly at the side of the road.

'It's not as tidy as Fair Isle,' said Simon.

'Apparently an islander called Ken abandoned them when they broke down.'

'Thank God he didn't have horses!' replied Simon.

The Isbisters must have seen us coming, as Bobby came to meet us outside the house. He pointed towards the reef, three miles away, and gave us some idea where he thought the *Oceanic* lay, but his directions were imprecise and of little use. We unrolled an Admiralty chart: the depth contours showed the northern peak of the reef where the ship had struck.

'We'll have to line up some landmarks on the island that are marked on the chart to position us on the reef,' I reasoned.

'Aye, we do that for fishing. They're called *meids*,' replied Bobby. 'As a boy I fished over the place she lay.'

'The reef's two and a half miles to the nearest land. We'll need big landmarks,' said Simon.

'We can sound the depth if we think we're near it,' I suggested.

The tide times were key to working the wreck. The tide was strong, quoted as up to 12 knots, and we would only be able to work as the direction of the tide changed, giving us a chance of slack water. Bobby did not have tide tables, so we would have to measure the tide on the pier to get the time of high water, and from that we would have to work out what it was likely to be on the reef. At least the *Oceanic* was an enormous ship, so the wreck and debris should cover a large area. The alternative was that she had slipped into water too deep for us to work.

Feeling that we had exhausted the subject with the Isbisters, Simon started on his friendly chitchat. 'The island must have changed since you were born?' he asked.

'The year I was born, 1899, that's the same as when the *Oceanic* was built. Two hundred and forty people on the island then, seventeen crofts on Da Hametoon growing grain and tatties. There are only five now, with just a few tatties and kale grown for the house.'

'Have you always worked the croft?'

'Ma faider built the house. I went to sea as a young lad, worked on sailing ships. The sea was the only way we could make a living off the island, many never came back, settling all round the world.'

'They're not keen to be dragged into the twentieth century,' Simon remarked after we left the house.

'No,' I replied, 'but they seem happy.'

'Perhaps life's happier when you don't have the option to make big choices.'

'It's as if everything merges here, the people are part of the nature of the island.'

'Rugged and isolated!'

'Perhaps,' I replied.

Up early, we had the inflatable in the water, loaded with all our equipment, in time for high water. On reaching the Shaalds, the water was flowing fiercely over the reef and the slight lift of the swell made it appear like a river in torrent. It looked unwelcoming and slightly frightening. Maybe the Isbisters had been right; perhaps it was undiveable?

We motored round to try and define the extent of the reef. It was easy to see the shallowest parts, as they had the strongest current flowing over them, the turbulence making it impossible to see the seabed below. We consoled ourselves that it might take some time to get used to the place. I was desperately disappointed but did not want to show it. We drifted away from the Shaalds, waiting for the current to ease.

When it eventually slowed, we returned to the reef and, except for a few swirls, the water looked more welcoming. Simon was keen to have a look and so slipped over the side with a coil of marker rope in his hand. He dived into oblivion, not sure what he would see on the bottom. A few searches were carried out only to discern a weed-covered rocky seabed.

Simon liked the searching, but I often found it difficult swimming a few feet above a seabed covered with kelp and oarweed. If there was a heavy swell, the weed would move backwards and forwards, and my brain would become disorientated. I often felt the seabed was moving and the weed was stationary. It gave an immediate feeling of nausea, the only cure being to hang onto the weed and focus on the stationary seabed. I never had this problem when working on an actual shipwreck, as my concentration was always on the wreckage, but Simon was unaffected by this motion so was good at searching. We stayed on site as long as we were able but found nothing. I recorded the slackest period of the tide, logging it as the first stage of making my own tide table. I was confident that we would be able to calculate each change of tide on the reef and from that work out our best time for working.

On our third day on the island our luck was in, as one of the islanders, Jim Gear, was out in his lobster boat and agreed to meet us on the Shaalds. Even with Jim indicating a position to dive, the first few efforts were unproductive. When the current started to increase, Simon would only have one more dive before it became too strong. The final position Jim suggested was too shallow, as we all believed the wreck had slipped off into deeper water, otherwise how could it disappear on one night within two weeks of striking the reef. I tried to haul the anchor up, but it was caught fast. Simon and I looked at each other, both hoping that the small anchor had snagged in wreckage. The current was increasing to a dangerous speed, so he went hand over hand down the anchor rope to prevent being swept away.

I thought of the vast propellers somewhere on the rocky seabed, the plates of steel scattered around them. In this shallow position it must have broken up, otherwise parts of it would be above the surface. My eye caught Jim's as we watched Simon's bubbles; they were coming up well astern of the inflatable, swept there by the tide as Simon would be ahead of us, circling around the anchor, hauling himself across the seabed, knowing if he let go he would be swept away by the current and have to surface a long way behind. I dipped my head below the water. The visibility was excellent, but I could only see a mass of weed, bent by the tide. I couldn't see Simon, as he was too far ahead. I checked his bubbles – they were rising further and further behind the boat. The tide was strengthening quickly. I expected him up at any moment. Concerned, I wondered if he had become caught on something, or if the tide was trapping him behind wreckage. What of the long scimitar-like steel spikes with razor-sharp edges that I had seen on the corroded double bottoms of a previous wreck? It was so easy to get cut by wreckage when you were thrown about by the current or swell. With a wetsuit it did not matter if it let in more water; it just became colder. I don't think either of us worried about ourselves when we were underwater; Simon always maintained that he was completely relaxed and I, though not always relaxed, was convinced I could handle any problem. Perhaps it was my youthful belief in immortality. I looked across the boat at my air bottles, ready to be used if necessary.

I felt tugs on the inflatable as Simon came up the anchor rope. Once on the surface, he held onto the grab ropes on the side of the boat, his legs pulled horizontal by the tide.

'I've found it, Alec! She's there, all broken up across the bottom,' he spluttered after spitting out his mouthpiece. 'The current's hellish. I can buoy her but haven't much time.'

Jim was holding his boat alongside. Simon passed him a coil of copper wire.

'Jim, it's for you: the first recovery off the wreck since the *Oceanic* sank.'

Jim took it and then Simon looked at me and held out his free hand, the other hand straining to hold him to the inflatable, the tide doing its best to tear him away.

'Here, Sy. Try and get it somewhere it won't chafe,' I said, as I passed him a prepared rope. Simon knew that any rope near wreckage would chafe on the sharp steel unless it was tied to the highest piece with a short length of wire rope attached to it. He hauled himself forward and disappeared down the anchor rope. Jim and I grinned at each other, elated by the news. I paid the marker rope out until I felt the sharp tugs from Simon that told me it was secure. Cutting off the surplus rope, I tied a buoy on the end before throwing it over the side. Leaning forward, I felt Simon tug the boat's anchor rope, as he required slack to free it from the seabed. After a few seconds he gave the command of two pulls to haul it in. I knew by the weight on the rope that he must be hanging on it, the boat drifting in the current as it was no longer tied. When he came to the surface I could see the anchor through the water with a bit of brass attached to it. I hauled it aboard while Simon held onto the grab rope. Taking the weight of his bottles, I let him slip over the side and lie like a stranded fish in the bottom.

'There's a bloody big ship down there!' gasped Simon.

Jim had an enormous smile on his face. I returned it as I slipped Simon's bottles off, allowing him to sit upright and tell us more.

'There's a hell of a lot of wreckage,' Simon said again. 'Large plates. It must be a big ship. It's got to be the *Oceanic*.' The tide had strengthened too much to go back into the water. I would have to wait for the next tide to dive.

We knew of no other large steamships on the reef, and it was unlikely another large ship could have been lost without someone knowing. Debris would have been washed ashore and most casualties were reported. I tried to persuade myself it was the *Oceanic*, as it seemed to be too good to be true. Simon had little more to say, other than that the seabed was littered with bits of brass. Like straining a teabag, I wanted to get every last bit of information out of him to imagine it myself. We motored back to the buoy, it would be pulled below the surface by the tide in the next few minutes and I had to look at the landmarks and line up *meids*. To the south-west it was easy: the gable end of the Isbisters house lined up with the cliff behind. To the north-west it was not so clear, and I took two bearings, the most prominent being a peat bank lining up with a dark patch in the cliffs.

On the way back to the pier we behaved like excited children, preparing to take the equipment off the boat, eyes meeting in smiles, ending with a thumbs-up. We said little because of the sound of the outboard, but inside I was screaming with joy, thinking money, salvage boats and a chance to stay on Foula.

After throwing our equipment ashore, we raced up to the Haa. I grabbed my diary and filled in the tide times, working out when we could dive again. It would not be today, as it would be dark at the next slack water. If we'd had a bottle of champagne, we would have opened it, but we made do with coffee. Our drinks finished, we ran up to tell Elizabeth, knowing it would be a matter of minutes before the news flew round the island.

That evening, as the initial euphoria melted away, reality began to creep in, along with doubts. We had achieved the first stage of our goal, to find a wreck, but we may not have found what we had been looking for. Had we truly located the *Oceanic*, or was it a different wreck? Would it be too dangerous to work effectively?

Would we get enough slack water to work it commercially? My brain was jumping everywhere. Were my dreams of making a fortune going to come true? Were there hundreds of tons of brass down there, just waiting to be lifted?

The next day's forecast predicted light winds. It should have been a perfect day but there was thick fog. We had no way of working in the fog because we relied on lining up landmarks to put us in position. A normal compass was inaccurate in a small inflatable because of the sudden movements of the boat. If we went out to the wreck, we would have little chance of finding it; the diver might get lost and we might not find the island on the way back. It was not a risk worth taking.

Part of the day was passed having fun with the two enormous Gear brothers, Ken and Jim, as they were preparing the mailboat for its next trip to Shetland. They were both incredibly strong and enjoyed testing themselves; one of them held our spring scales above his head, while Simon or I clung on, the other calling out the weight. It took our mind off the wreck . . . if only briefly.

It was several days before the weather cleared and we were able to revisit the wreck. I dived first and lost no time in finding as much scrap as possible: copper and brass in the form of pipes, valves and small fittings of every sort, from lights to portholes. I had little time to spend exploring her, as our priority was to recover scrap, but when I swam past a boiler I saw the vast engines towering behind it in the clear Shetland water. I now knew without doubt that we had discovered the wreck of the *Oceanic*. These were the enormous pieces of machinery that had pushed her across the Atlantic on her weekly voyages, the stokers shovelling coal up to the rate of 700 tons a day into her massive boilers. The machinery was unmistakable; I had no doubts.

When out of the shelter of the engines, I felt the current pouring round them like a river and worked my way along one of the propeller shafts, hand over hand on bits of wreckage, the current tearing at my body, attempting to break my hold on the wreck. As the strengthening current became too much, I reached out to grip the remains of the hull. Immediately I felt pain in my hand, as the sharp steel cut through my glove. Instinctively I let go and was whipped away, tumbling over the wreckage. I headed for the surface, where I waited for Simon to release the inflatable and pick me up about a quarter of a mile from our mooring.

'What d' you think, Alec?' he asked when I was only half in the boat.

'It's got to be the *Oceanic*, Sy. After a brief look at the enormous engines I have no doubt they are from the largest ship in the world.'

'D' you think we can work out here?'

'As long as we get a period of slack water it doesn't matter if the current runs up to 12 knots. At least we'll never work long enough to worry about the bends.'

'We could take the easiest bits and go,' answered Simon.

'No, if we do it well, and watch our costs, it'll set us up for life.'

We always took it in turns to dive. One of us would attach the brass on the seabed and the other hauled the bits into the inflatable. By the end of a short twenty- to sixty-minute working period – all the tide allowed – we had nearly half a ton, with a value of between £400 and £500. The inflatable was grossly overloaded: the diving bottles were perched on top of the scrap, with the two of us holding on as best as we could, the water slopping over the almost-submerged pontoons as we made our way slowly back to the pier, elated with our recoveries.

The more scrap we could recover, the stronger our position would be as Salvor in Possession, a legal term used when the recovered material was declared to the Receiver of Wreck, declaring it initially as coming from 'a wreck or wrecks on the Shaalds of Foula' to cover ourselves if it was not from the *Oceanic*. We phoned Mr Rowden at the Salvage Association to find out if we could trace the owner, but he said the wreck had been sold in 1924 and he had no knowledge of who had bought it.

Ownership of a wreck can be complicated, as there are normally three insurance companies involved with a ship. The hull underwriters, who insure the hull and machinery, the cargo underwriters, who insure any cargo she has aboard, and the Protection and Indemnity Clubs that insure the liability of the ship in the event of a collision or oil spill. Each of the insurers are made up of hundreds of names entered on their books who are committed to pledging their money towards any payout. In some cases the insurers have no interest in taking on the ownership of the hull; even if they pay out, the owners retain ownership and keep the ability to sell it to someone like us. Because we knew the curator of the Lerwick Museum and we'd been to see him to discuss what he would like us to do if we found a wreck of historic interest, when we phoned him he informed us that a company called Robertson's owned the *Oceanic*. (As an afterthought, when the call was finishing, he asked us to chisel a piece of rock from the Shaalds, as the geologists would be interested and he required a sample for the museum. It had to be chiselled, he insisted, not loose rock, as he had to be absolutely sure where the rock came from.)

Further research showed that Robertson's had been active in marine salvage in the 1920s but had moved their core business to the role of ships agents and held the BP fuel agency. Most of their work was in handling the affairs of Icelandic, Faroese and Danish fishing vessels, although the much larger Hay & Company in Lerwick was the main Shetland agent, as well as being ship owners themselves. They decided to buy Robertson's in 1969, thereby unintentionally inheriting ownership of the *Oceanic*.

Because they were not involved in salvage, it was easy for them to assume that a wreck was like other property and they could do what they wanted with it.

After phoning Hay & Company and not receiving a healthy reception without knowing why, I thought it would be best to have a face-to-face meeting next time we were in Shetland. Before that, the most important job was to continue making recoveries to strengthen our position. The pile of scrap grew and, with the difficult conditions we worked in, the strong tide and heavy swell, we both had a tremendous sense of achievement. We no longer had any doubt that it was the *Oceanic*, although we had not recovered anything specific to identify her. We regularly phoned the Receiver of Wreck in Shetland, John Butler, the head customs officer, now declaring our recoveries as from the *Oceanic* prior to him sending us the official declaration form.

Worryingly, several days after declaring our recoveries, we started receiving phone messages from a Glasgow company, the agents for Mr Arthur Nundy, who ran a large salvage and underwater engineering company. We worried that they might turn up on the wreck, as they were making us miserly offers on our salvage recoveries, claiming they had an exclusive contract on the wreck. Concerned, but not sure about their authenticity, we gave polite but non-specific replies, stating that we could not answer them unless we saw their contract and explaining that it would be two weeks before the mailboat went to Shetland and collected the mail. The only people the Glasgow business could have a contract with was Hay & Company as the owners of the wreck and we thought that might explain why Hay & Company were reluctant to listen to us.

The wreck was like nothing I had ever seen. Initially we had no intention of exploring it: there was plenty of brass to lift beneath our first mooring. But once we had recovered enough material for us to be certain we could firmly establish our position of Salvor in Possession, we began swimming further away, spending a short time each tide exploring before the tide strengthened and lifting became difficult.

The seabed was rock, covered with kelp and oarweed that waved backwards and forwards in the swell, except when the current was strong, when it would bend over like bushes in a strong wind, lying almost flat against the wreckage. It was very different from the wrecks I had seen on Barra, where there were pools of sand in concave areas of the seabed. On Fair Isle very little of the steel ships remained, but here there were numerous steel plates that would have made up the sides and deck of the ship, reflecting the enormous size and its heavy construction. Shiny places were visible where the plates had moved in bad weather and rubbed together, scraping off the weed and their coating of rust. Some of the rock was broken and lay where sand might have lain if the site were not so exposed. In other areas there were cliffs with wreckage piled up against them. The visibility was good, but not as good as Barra, and I often received a sudden fright when I noticed an enormous object towering over me. In most areas it was possible to work out where I was on the ship by the parts that remained on the seabed, but we had to be careful, as some of the fifteen vast boilers had worked themselves loose and rolled away from their original position. When I eventually explored the engines in detail, I could see their massive weight had kept the double bottoms of the ship in place beneath them. It gave me an indication of the size and heading of the wreck. The engines were thirty-five feet apart and towered above me, and the port engine had fallen over to lean on its neighbour.

The two steam engines were almost the same in size as the two outer engines on the *Titanic*, although she had an extra central turbine to supply her additional power requirement. The engines were described as triple expansion engines, although they had four cylinders, as two intermediate cylinders were required to cope with the increased volume of steam produced as the pressure dropped and the steam expanded after it had pushed the first piston. The fifteen enormous boilers and engines required 200 men in the engineers' department to keep them going. Once the steam was exhausted from the engines it was sucked through the steam condenser that cooled it back to fresh water. Each of the four condensers had about 3,000 brass tubes and the *Oceanic's* were unusual in that they all had brass casings, each of them weighing about seventeen tons. I could hardly believe my eyes.

I swam along the port engine crankshaft, the part of the engine that turned the up-and-down movement of the pistons to a rotary motion. It weighed 110 tons and had vast bearings on it to allow it to turn. By scraping the weed I could measure the thickness of the brass bearing shells that were held within cast-steel housings. They were enormous; each complete shell on every journal weighed over half a ton of gunmetal, a top-quality brass that contained tin. On their surface they had 200 lbs of white metal that was worth £5,100 a ton. Tons of copper pipe, valves and other items lay loose among the engines, torn off by the heavy seas. Looking aft along the propeller shafts, I calculated there were twelve horseshoe-shaped thrust blocks on each shaft, used to absorb the 28,000-horsepower push of the ship's propellers and transfer it to the hull. Without them the engines would have been destroyed. These blocks contained tons of white metal.

The propeller shafts were twenty-four inches in diameter and there were nine sections to reach the propeller 225 feet away at the stern. Each section weighed twenty-four tons. Every revolution of the two twenty-nine-ton brass propellers would have pushed the ship forward a distance of thirty feet, and from past experience on wrecks there was a chance that spare propellers or blades were stowed near the stern. The *Oceanic's* propellers were just six inches smaller in diameter than the *Titanic's* two outer propellers.

The sides and deck of the ship had all collapsed. It could not have been better for us – all the parts we wanted were exposed, making them accessible to blast off and remove. Remains of the steel plates lay flat on the seabed all around the ship, disappearing into deeper water where we did not wish to go. Only the portholes within them were of value and some of them had broken off due to corrosion and heavy seas; they lay on the seabed waiting to be picked up.

It took a good number of working days before I had the time to swim as far as the propellers. On the way I saw five spare blades at six tons each, before finding the two main propellers aft. One was partially buried in wreckage and debris, while most of the other was exposed, the top blade broken off the starboard propeller and lying clear, waiting to be lifted. The enormous rudder lay on the starboard side; its position made it look as though it had knocked the propeller blade off when it collapsed. The three-bladed propellers would require some serious blasting to cut them off the shafts and make them small enough to lift aboard the size of boat that we could work from Foula.

When Simon hauled me into the boat, I was bursting to tell him what I had seen. 'It's fantastic, Sy! I've never seen anything like it! Several hundred tons of brass to recover. Much of the bigger bits require blasting free.'

'How do we lift it?'

'It should be easy,' I said, my confidence born from my excitement.

'What about loose brass?'

'There's still tons of loose bits, but we'll need a proper boat for the big bits.' I knew he was not keen on a salvage boat with all the problems it would bring, such as keeping it at Foula.

'Let's just get what we can,' said Simon, echoing my happiness.

'I'll sketch it out for you when we get ashore,' I said, not wanting to start a discussion.

Simon knew little about engineering – and had even been
known to let three girls change the wheel on his car when he'd
had a puncture – but that was probably Simon being Simon.
His lack of knowledge did not stop him recognising the various
metals underwater and slinging them to lift, but I knew I had to
persuade him we required a decent salvage boat.

While we were working on the reef, Harry from the post office
kept an eye on us, as he was worried for our safety. He had even
mounted an old First World War gunsight on top of a fence post.
His monitoring of our work also meant he knew when to meet
us, giving us a hand to lift our recoveries ashore. He looked at
each piece, trying to work out where it came from, fascinated
by how such a large ship could have broken up into such small
pieces. Sometimes he would be joined by Elizabeth and other
islanders curious to see what we were recovering – and probably
seeing little purpose in it. Hovering in the background was always
Tom, pipe in mouth, waiting to catch us on our own to tell us
about the weather or some island story.

Of the population of twenty-eight people there were only four
men and five women under the age of forty-five permanently
on the island. Several months before our arrival, two marriages
had broken down, resulting in the loss of three young people.
It was hard financially for the few young to look after their own
families, as well as the older people. Being given a hand with small
jobs could make a big difference to them. Simon mentioned to
me that Elizabeth had a growing opinion of our usefulness; he
claimed it had come to a head when we were walking up to her
croft and found the road blocked by an old grey petrol paraffin
Ferguson tractor. Elizabeth's brother was on the island for a few
days and was trying unsuccessfully to reverse a trailer load of peats
into her yard. Simon egged me on to do it for him, but I felt it
might be insulting to offer help, though I changed my mind when

Dan, Isaac, John, Peter and Alec on the *Vesper*, which we used to salvage the *Salvestria*. (James Price)

Above. The *Dewy Rose* at Northbay, Barra.

Left. Peter preparing sacks of scallops on Castlebay pier. It was he who persuaded me to dive on Barra.

Alec and Chris studying the *Adelaar*'s cannon, recovered off Barra on an archaeological project. It was while working with Chris that I met Simon.

The *Dewy Rose*, working on the wreck of the *Canadia*, Fair Isle. On this attempt we failed to lift the propeller.

The factory manager and Simon with our morning's haul of scallops, Grimsay, North Uist.

Left. Chris with one of the many whisky bottles we uncovered on the wreck of the SS *Politician*, Eriskay.

Below: Our Haflinger travels to Shetland, where Simon and I would spend the summer searching for wrecks.

The mailboat *Good Shepherd 2*, which made the journey twice a week from Fair Isle to Shetland.

Above. The team – Alec, Simon and John-Andrew – preparing to get into the inflatable.

Right. Elizabeth, our friendly next door neighbour who arranged our rent of the Haa on the island of Foula. (Simon Martin collection)

The *Oceanic* at Belfast.

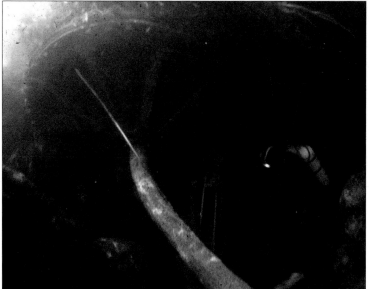

Above. A bearing on one of the 110-ton *Oceanic* crankshafts.

Left. Working in among the *Oceanic* engines.

The *Good Shepherd 3* being loaded with our first scrap recoveries at Foula. (Simon Martin collection)

One of the *Oceanic*'s anchors used as a mooring for *Trygg* at the entrance to Ham Voe. The massive chain had to be cut with explosives.

Trygg at the entrance to Ham Voe.

Above. The remains of *Trygg* floating inside the pier after the storm.

Left. *Valorous* anchored over the *Canadia*, as we finally recovered the remains of her propeller.
(Colin Martin)

Above. Moya relaxes on *Valorous* after operating the winch. The beam running along the centre of the hatch is the one that fell on me.

Right. At seven and a half tons, this is the smaller deck removal grab that we both used and marketed when the family business progressed. We designed them to work at full ocean depth.

I realised he was stuck. Her brother could not get off the tractor fast enough when, slightly nervous that I had forgotten the skill, I told him I could back the trailer to where Elizabeth wanted it. Simon claimed it was then that he'd spotted a gleam in her eyes, as if she'd won the jackpot.

Following this, we were able to return Elizabeth's kindness by being trusted to mend fences, catch sheep and undertake mechanical repair work on her generator during bad weather days. It was a loose barter system and as word spread round we often worked at other houses with an elderly person. It became an enjoyable change, taking our minds off the wreck and offering us a real way of attempting to repay people for their help. Elizabeth became more and more relaxed, telling Simon gossip she should not have been telling, and smiling when she met us before asking, 'How are my divers today?'

We quickly got to know everyone and I was beginning to wonder which I would miss most if we were thrown off the wreck by this Glasgow company – the island or the *Oceanic*?

A small 32-foot fishing boat called the *Lustre* became a regular visitor to the Foula pier. There was only one man on it, Kenny, who was trawling off Foula during the night with little success. The boat was just too small, but it looked as though his small derrick for lifting in the trawl could take about half a ton and the boat would carry a few tons of scrap. Seeing this as an opportunity to lift larger pieces of scrap, we asked if he would hire his boat for a tide or two on the Shaalds. He jumped at the opportunity. Simon and I would both dive and, requiring another hand, we asked Ken Gear to help us too.

Full of enthusiasm as we left the pier, Simon and I in the inflatable and the two Kens in the *Lustre*, we were in luck, as it occurred at a time of 'neap' (slack) tides. These occur seven days after a spring tide, when the sun and moon are at right angles to each other. Our time would still be limited, but we would get over an hour's working time.

Arriving on the Shaalds, we tied the inflatable behind the *Lustre* and I dived, taking the end of a rope from the boat. Knowing exactly where I was going, I headed for a pile of three spare engine bearings, each half bearing weighing about 800 lbs. Tying the light rope to the first one, I gave the single pull for attention. Simon responded and I could feel him taking the rope in, as the *Lustre* manoeuvred over the position, the noise of the boat's engine increasing above me as she came closer. A shackle attached to the lifting wire was slid down the signal rope and I quickly attached it to the bearing, following it with a single pull on the rope. I waited for Simon's single pull in reply before I gave him two as a sign to lift. Within seconds the bearing was plucked from the seabed. This was easy work and the other two followed in quick succession. Moving to other lifts, we quickly filled the *Lustre* before the current stopped us. Proud and exhilarated by our work, we unloaded the brass on the Foula pier. They were the first properly recognisable pieces that we'd recovered and could easily be identified as specific parts of the wreck of the *Oceanic*.

Ken and Jim, the two enormous Gear brothers, were having a look at trying to lift one of the bearings. 'If you can lift one of those with one hand, you can have it,' Simon said. 'Mind, one hand only!' he shouted, worried that one of them might actually lift it!

The following day we all went out again. Simon dived first and then I took over as he held the signal rope on the scrap-cluttered deck. The current was starting to increase, but we were all enthusiastic, feeling the job was easy. Securing a large boiler valve that lay beside one of the enormous boilers, I gave the hoist signal. I saw the tension come on the wire, but the rope was at an angle, not pulling vertically. I quickly swam out of the way as the valve bounced across the seabed before catching on some wreckage. On deck Ken was pulling hard on the winch, but the current had started to flow and Kenny in the wheelhouse was no longer able to keep the boat's head into the current. She swung round broadside to the current effectively moored by the top of the derrick. As she leaned over, the engine stopped with the load, the water rising up one side of her scrap-filled deck. I had swum to the surface and saw the boat almost on her side before Ken was able to release the winch and let the wire run out. With the wire released the *Lustre* and inflatable drifted off, Simon jumping into the inflatable to come and pick me up.

'We nearly joined you,' he said jokingly, as he hauled me into the boat.

We did not risk using the *Lustre* again, but it was a good lesson on how we should have planned our lifts better. I made notes in my diary on a method of working in the future, if we bought our own boat. Like working with Chris, I was learning all the time on the *Oceanic*. I realised how much I still had to learn.

With heavy seas increasing through autumn, our diving time became limited. We found out how dangerous the Shaalds could be when the weather quickly changed. After a gale we would be lucky if our mooring on the wreck remained. It seemed almost impossible to prevent it from being cut on the wreckage, as it would drop down in the hollow of the swell and snag under steel plates, pulling the buoy below the surface until the buoy was damaged or its tether was cut on the sharp wreckage. But we had become good at using the *meids*, lining up prominent places on the island, and I would position the boat before Simon went in to lay a replacement mooring. If the visibility was good, I could drop him on top of the engines.

Knowing our limitations with the current, we just worked until we couldn't work any more, but the heavy swell was difficult to judge and it was dangerous. The inflatable lay nicely on the mooring if the swell was not breaking, but beneath the waves the motion throwing the diver backwards and forwards was so great that it would tear our grip from anything we held on to. Wear and tear on our suits, which were cut regularly, sometimes right through to our skin, became so bad it deterred us from working in marginal weather. Otherwise our equipment worked well. And apart from one minor fault with the outboard at sea, we had no breakdowns.

The Foula mailboat held the contract with the post office to deliver the mail on one round trip every week in the summer and once a fortnight in the winter. There were scheduled times posted in the village of Walls on the Shetland mainland. In reality, it was totally weather-dependent and the boat rarely sailed at the allotted times or on the allotted days. Jim and Rob, the crew, were always cautious about the weather conditions in which the boat could safely make the trip. As a consequence, it was usually after 10 a.m. before they sailed – much later than the time on the schedule. The boat's old Perkins engine required warm air to be sucked into the air intake before it had any chance of starting. This was carried out by either a small gas blowtorch, or, if Jim had run out of gas, old paper was burnt next to the inlet manifold, allowing the warm air to be drawn in. When black smoke poured out from the exhaust, there was a sigh of relief, as we knew the engine had started. The sole method of navigating the boat was a compass, and she had no radio communication on the three-and-a-half-hour trip to Walls. This boat was the mainstay of the island; a vital service which enabled the island to survive. On other islands people had been united by the church or a community hall as well as the boat. On Foula, it was only the boat that brought people together.

When the stores were unloaded, some were taken up to the shop, which was housed in a two-storey, slate-roofed building that was in desperate need of repair. It was the nearest building to the Haa. We felt it a duty to use the shop, but there was little food in stock, and most of it was tinned. Harry was running it temporarily, as the previous shopkeeper had left the island. In better days the shop would have opened when the mailboat came in, the islanders gathering to help unload the boat and carry the fresh stock up to the shop before buying it. Harry being Harry was helping out – the island opinion of him was that he had never been heard to say a bad word about anyone and was, therefore, the most respected man, as well as the most trustworthy. From our short time on Foula, I had seen that he was a gentle man of deeds and considered any situation carefully before using words. I believe he was the glue that held the island together, having an equable solution for most problems.

It was a wet day when we first went into the shop. Harry was standing behind the counter with his oilskins on, buckets placed strategically around the floor to catch leaking rainwater. Boxes of tinned food were stacked in dry areas, with only a few groups of tins placed on the shelves. The cardboard boxes looked damp, but the tins appeared free of rust, and those in plastic wrapping were like new. Few of the tins had been taken out, as they were normally sold by the case. Simon and I looked at the labels. With the stock run down, the selection was poor. The Carnation milk and corned beef that we consumed regularly had sold out, so we bought a few boxes of beans and tinned fruit, but more from goodwill than immediate necessity. I paid Harry, who rummaged through a wad of notes to give us change. The notes were a strange colour and looked as though they were from a long defunct bank.

'They look like old notes, Harry,' I remarked, as I flicked through the change, not to count it but to examine it.

'Boy, boy, I'll sort newer ones for you,' he said.

'No, they're okay. I like them.' I smiled as I put the change in my pocket and we lifted three cases each to carry them to the Haa.

'I'll bring the last two after I've closed up,' said Harry.

Outside the shop the rain was moderating and the sun was coming through. I dumped the boxes on our kitchen table and took the notes out of my pocket. There were four. I gave two to Simon and looked at the other two.

'It's a 1933 Commercial Bank of Scotland £1 note; the other's a British linen banknote,' I told him.

'I've got a '37 Commercial Bank and a Union Bank of Scotland note,' he replied. 'Do you think they're worth anything?'

'I don't suppose so. They're not in very good condition. They've been handled too many times.' I bent back the dog ears to see if that improved them. 'They'll never look like new.'

'The only banks on Foula are peat banks,' said Simon, as he took the notes and put them in the tin beside the petty cash book.

As the days passed we became frustrated waiting for the mailboat to go out and bring back the forms the Receiver of Wreck had sent. We had filled them in for other wrecks and knew one of the questions would be the value of recoveries. We were now certain that Nundy had a salvage contract with Hay & Company, the owner of the wreck, and we wondered how Nundy and his agents would react: if we put in a low value, it might put them off the wreck, but it would also show we were not serious salvors. A high figure would show us to be good salvors but make them keener to get rid of us. We settled for a true figure, weighing the scrap, dividing it in its different categories of metals and phoning a metal merchant to obtain current prices. The value exceeded the cost of a three-bedroomed Victorian stone-built house that a friend of mine had recently bought in a village near my parents – and the excess above that would cover all our costs. This was after less than two months' work, and with only a 13-foot 6-inch rubber inflatable boat!

'Are you sure your figures are right?' queried Simon, hardly able to contain his joy.

'That's the third time I've checked it. You have a look, Sy.'

I was thinking how much more we would make if we had a larger boat, but knew it was not the time to broach the subject or the time to buy a boat before winter.

The weather continued on the borderline for work on the wreck and we had to stop when it worsened to heavy rain, driven by gales.

Relaxed and enjoying the warmth of the peat fire in the Haa, we heard Elizabeth banging on the door. 'Boys, boys, yon Glasgow man has phoned about yon wreck again,' she shouted through the half-opened door. Elizabeth stood in her oilskin jacket, a skirt just below the knee, heavy stockings and the usual pair of wellingtons. Water ran off her jacket as she held it down to stop the wind getting under it.

'Are you coming in?' I offered.

'Naw, naw, it was just to let doos ken,' she said, turning round to go back to her croft.

'Thanks. We'll look up later,' I said before closing the door.

Simon had remained in the sitting room. He had damped down the fire with broken peats and was starting to put waterproofs on. 'We'd better make this call from the box and keep it to ourselves.'

We regularly discussed what we thought would happen next. We knew Nundy was trying to oust us from the wreck by claiming he had sole rights to the salvage. He had not called for a while and we had hoped he had gone away, but now I knew it was a matter of time before it would end up in court. We were damned if we would give the *Oceanic* up to a company that had not even attempted to dive on the dangerous Shaalds.

6

'Salvage Men in Legal Battle'

We struggled against the force of the wind and rain to reach the phone box. Simon entered first, shifting the opened bag of hen's layer pellets to make more room. I pulled the door firmly against the wind until at the last moment it slammed shut. Stamping my feet and shaking my body, I watched a pool of water grow on the concrete floor. We referred to the phone box as our *office* – we used it solely for that purpose, although it was normally treated by the islanders as a general store because it was dry and convenient. I had even seen a lame sheep and some motherless lambs shut in to be tended to later. There was no light inside, though the small panes of glass let in sufficient daylight.

'Have you got the number?' asked Simon, as he picked up the black Bakelite phone and wiped it with his hand. I removed wet gloves with my teeth and opened my waterproof coat to search the inside pocket for my diary. The space was cramped and my cold fingers fumbled with the pages, making damp splodges on the writing. Simon filled the box with coins, as I read the numbers out for him to dial. Our breath came like smoke and condensed on the small glass panes. I fidgeted nervously, rubbing at the condensation with my gloves for no real reason. The rain dribbled down the inside, where some of the putty was missing. I wiped it with my elbow, attempting to remove the oily streaks left there by my dirty gloves. Without some repairs, this place would not be watertight much longer.

Hearing the ringtone at the other end, I stood motionless with my heart in my throat. Simon glanced at me with a look of disappointment. Slipping my diary back into my inside pocket, I zipped up my coat and started to put my gloves on, getting ready to leave. Then the ringtone stopped. I heard the faint click of the phone being answered. Simon pressed the 'A' button and we heard the coins fall.

'Good afternoon, this is Nundy's agent,' came a business-like voice.

'It's Simon Martin here, returning Mr Nundy's call.'

'I'll put you through.'

Simon smiled at me with his hand over the phone. I leaned my head towards him to hear the conversation, my cap pushed askew as it came up against his sodden bobble hat. The noise from the wind battering the phone box made hearing difficult.

'Hello, Mr Martin,' the stranger answered before getting straight to the point. 'Have you considered our offer?' I tilted my head back. Simon was smiling, a cunning smile; our eyes met in total agreement. He gave a very slight nod.

'We will neither accept nor refuse your offer,' Simon said and looked at me. I moved my head back towards the receiver and took my cap off, as it was getting in the way again. The voice at the other end of the phone changed. It became angry.

'You mean you refuse the offer?' he shouted. I moved my head away, knowing I would now be able to hear without being close to the phone.

'No,' replied Simon. 'We neither accept nor refuse the offer,' he repeated calmly.

'You can't. You can't. You have to accept or refuse,' came

the voice, which had increased so much in volume that Simon emphasised it by moving the receiver away from his ear.

I put my cap back on, excited that our ploy was working. This would give us the few days' delay we required.

'We are neither accepting nor refusing the offer,' replied Simon, quietly adding, 'as you have refused to let us see your contract.'

I fed the phone with more coins, hoping that the agent could not hear them drop into the box. Simon moved his hand to indicate a throat being cut. I nodded.

'If there is nothing else to discuss, we have other business to attend to.' Simon then repeated our *office* number, stressing that it was best to leave a message at Elizabeth's to check whether we were out at sea.

'You will be hearing from us,' came the jaded response, now with a softer, more thoughtful tone.

'Thank you,' replied Simon, as he replaced the receiver.

We looked at each other, cold and wet, jammed together in a phone box, moist air pouring off us, pretending to be something we were not. We had nearly been beaten, but now we were full of hope. 'I think it worked, Sy,' I said. 'It's given us the time we need. It's confused the issue.'

'Yes,' Simon replied. 'Let's go to Elizabeth's and celebrate with a cup of coffee.'

Squeezing out of the phone box, with a tight hold on the door in case it blew off its hinges, we left the *office* and, bowing our heads into the horizontal rain, we followed the track to Elizabeth's at a slow run.

Outside the door was her pet lamb, Pizzi-izzy-ewe. It had contempt for the fencing and had broken into her yard again. Well past the age of being a pet, it still expected to be fed and it was always a struggle to keep it out of the house. It had had months of training, forcing people out of its way in an attempt to get at the food in the porch – often succeeding.

'Oot, oot, oot!' shouted Elizabeth, opening the door. She tried to push the lamb away. 'Come away in, boys,' she said, as Simon and I slipped in behind her. The door shut with a bang, leaving a wet and angry Pizzi-izzy-ewe outside. 'How'd you get on, boys?' she asked.

'I think we've got them rattled,' said Simon, accepting a cup of coffee. Sitting beside the Rayburn, old Joann placed the shawl she was making on the resting chair. Slowly putting a hand to her head, she moved her fine grey hair away from her face, gave a gentle smile, a nod and a faint 'hello'. I noticed how much the shawl had grown during the week. Simon continued in his upbeat manner in the knowledge that Elizabeth would absorb every scrap of information, almost certainly relaying it around the island after we left. She had been kind to us and deserved to hear any details first, but we had little real news to tell her. As we were happy, I suspected she thought we'd had a better outcome than our phone call warranted.

By the time we left to walk the few hundred yards to the Haa, the wind and rain had moderated. Low dark clouds still threatened, but it looked as though the worst had passed. Entering the house I threw my oilskin on the back of a wooden chair and started to light the Tilley lamp, my thoughts returning to the phone call. Simon prodded the smouldering peat until the flames burst out, then nestled the kettle into the hottest part. The house felt cold compared to Elizabeth's. It would take a while to warm up.

As the light from the Tilley gradually spilled over the room, I realised we had been having so much fun salvaging the *Oceanic* we'd managed to put to one side the nagging doubts about our right to work it. We'd had regular phone calls from Nundy's agent, threatening us with the police and claiming ownership of our salvage. Although we had always answered the calls, we had rather hoped it would all go away without us doing anything. We had spoken to the Receiver of Wreck and the owners, Hay & Company, and believed we were not doing anything illegal.

'What d' you think they'll do?' I asked Simon. 'Will they take us to court, do you think?'

'It's the only way they can stop us. It's a civil action, not criminal. It was just bluster when they originally threatened us with the police.'

'It's the money,' I said. 'They'll know how much it's worth.'

'They've nothing to lose. We have,' replied Simon. 'I hope we've delayed them enough to strengthen our position.'

I was beginning to fall in love with the island, looking at old crofts with the thought of renovation. I knew Simon would have preferred the *Oceanic* to have been wrecked off Fair Isle, but he was happy enough on Foula. The two islands were so different, it was as if they reflected the difference in our two characters: he preferred the ordered life, whereas I preferred the potential resulting from a lack of order. Although Simon had intended our summer on Fair Isle to be his last salvage work before returning to a 'proper' job, it was now obvious that he was as committed to the *Oceanic* as I was. The more Nundy tried to get us off the wreck, the more we dug our heels in. Taking all the risks, we were damned if someone would wrestle it out of our hands. We knew the law and had followed it by informing the Receiver of Wreck and claiming the legal term of Salvor in Possession.

'Their only argument is that we're not competent as salvors,' said Simon, hanging his soaking bobble hat on the brass rail above the peat fire. 'If past court cases stand, it will be the same as in the book.'

I had the court statement in an old salvage book that lay open on the table. We had been looking at it before we went up to make the call. It read: 'In the case of a derelict the salvors who first take possession have not only a maritime lien on the ship, but they have the entire and absolute control and possession of the vessel, and no one can interfere with them except in the case of manifest incompetence.'

'We need to get our remaining salvage equipment from Fair Isle,' I suggested, closing the book. 'If we had the stuff, life would be easier. It'd also show a total commitment to the project.'

'It'll take weeks. We've got to take the Foula mailboat to Shetland, go to Grutness, catch the Fair Isle mailboat and the same on the return.' Simon sighed. I knew he was thinking about the Foula mailboat, which at best was only running every two weeks. He walked across to his usual place on the settee, picked up an old magazine and started to thumb through the sports pages, as if settling down for the evening.

'Why don't we nip over to Fair Isle with the inflatable and pick up the gear? We'd get the Fair Isle mailboat, but we can almost guarantee the days it will run,' I said, excited at the thought of the crossing.

Simon put the magazine down. 'Our 13-foot, 6-inch inflatable, on a 50-mile open-sea crossing? It's a bit of a risk. We'd have to make sure the weather's good.'

'If the swell's behind us and there's good visibility, we'll be okay. We've two outboards and plenty of fuel. It's worth a try.'

'It's further than going to St Kilda from the Western Isles.'

'There's always a few fishing boats about, Sy,' I assured him.

'When d' you think we should leave?' asked Simon, edging towards the idea.

'No later than Sunday. That gives two days for the weather to improve.'

Simon nodded as he picked up the magazine again, put his feet on the settee and returned to the sports pages.

Returning from Fair Isle five days later, and leaving our equipment at Walls, we used the inflatable for the last part of the journey between Shetland and Foula. The mailboat would return the following day, as we had seen it in Walls and spoken to Jim about transporting all our heavy gear. As predicted, it arrived late at night, but it was the following morning before Harry sorted the mail and brought us a registered document from the court that contained an injunction preventing us from working the *Oceanic*.

Nundy had made three applications to the sheriff within the one document. The first was a full interdict restraining us from removing, selling or interfering with any part of the *Oceanic*, or from selling, removing or interfering with any part of the *Oceanic* already salvaged. The second, as an alternative, was to grant a similar interdict to the first, but an interim version, and the third asked that we be ordered to hand over to him any bits from the *Oceanic* we had already recovered. This we considered was not legally possible, as we had declared all our recoveries to the Receiver of Wreck and technically they were in his custody, although they were kept on Foula. The sheriff granted the second version or interim interdict, which prevented us from working the wreck, but he also called for us to reply. In the papers, Nundy outlined the contract that he held with Hay & Company, which gave him 'exclusive rights' to recover and sell material from the wreck.

There was nothing else we could talk about. We discussed our options: either to fight the case or give up, with the possibility of negotiating a reward on the material we had recovered. Initially Simon had been swithering, as he suggested recovering as much as we could and then leaving, but now, as a new challenge had been presented, he was determined to pursue the legal side. *We had to fight it*. Our delaying tactics were over. It was time to hire a lawyer.

We immediately set off in the inflatable to make the crossing to Walls, taking PI's taxi on the three-quarter-hour trip to Lerwick, only just managing to catch the steamer south. It was

6 a.m. on Sunday morning when we arrived in Aberdeen harbour and caught a train to Fife. We now knew that Nundy had sent a diving team to try and find the *Oceanic* before we'd arrived on Foula. They had been taken out to the Shaalds but decided not to risk diving in such treacherous waters. Our weak point was our competence: Nundy owned a large, well-known salvage company. But while we may have been two scruffy divers, we had dived on the *Oceanic*.

We brought some of my books citing old salvage cases to show our lawyer in Edinburgh, assuming he was new to salvage, but he had already researched the topic and was up to speed with some of the previous cases. He advised us to put the following argument to the court:

> The wreck of the *Oceanic* having lain on the seabed abandoned for many years falls to be classed as derelict, with the result that any person is entitled, irrespective of contract, but subject to due observance of rights which may have accrued to other salvors by virtue of prior possession, to dive and locate a wreck and to recover parts of the vessel or cargo.

We admitted diving on the wreck and refusing to stop work when asked, claimed that we were exclusively in possession and that Nundy's action was an attempt, without just cause, to interfere with our rights as salvors. Leaving the lawyer, our confidence was slightly diminished, as he reminded us that our competence might be questioned at a later date and we should prepare for it by obtaining references from past employers.

Simon and I had looked at each other.

'Chris could end up in jail if he declared the brass he'd taken,' I said.

'He won't want to expose himself to scrutiny,' added Simon.

'I'm not sure Dan would even want to do it. He had told me his uncle was intending to go into a joint venture with Nundy's salvage company on a wreck in Scapa Flow. We're just going to have to hope it doesn't come up. What d' you think they'll do, Sy?'

'Don't know. We could be in for a pasting if Nundy's company is compared to anything we've worked on,' Simon replied. I knew exactly what he meant.

'We'll start to run up big legal bills unless we're careful.' I thought of all my savings disappearing and realised there was nothing on Foula for us without the wreck. I might have to start again, but where and with whom? I enjoyed working with Simon, but I knew he would give up diving if we lost the court case.

Simon broke across my thoughts. 'It's too good a chance for them to miss. D' you think a big company would make money on it?'

'Their costs'll be high. They'll lift the big bits and leave the rest, but they could screw us as a matter of principle,' I replied.

'And I thought this was our big break,' Simon sighed, disheartened.

As we wandered down to Haymarket station from the posh lawyers' office in Charlotte Square, I envisaged my life falling apart. We'd been so full of hope when we'd entered the office. By the time we reached the railway station, we were both pretty glum.

'We've got to base our competence on our work on the *Oceanic*,' I said to Simon. 'It's not our past competence that's to be judged, we just have to show we're not *manifestly incompetent* on the *Oceanic*.'

'Let's produce a document. If you produce some drawings, I'll do the text,' said Simon, as we discussed the proposal on the train journey to Fife.

By the time the train arrived at Leuchars, we had worked out a plan and at my parents' home Simon phoned Elizabeth. We were both concerned that if the weather improved the opposition might attempt to dive and make recoveries that would further confuse the situation. Fortunately the weather had remained bad, and the three-way conversation with Elizabeth cheered us up; there was no way we were going to give her any hint of losing the case, as she had absolute faith in us winning.

'I'll stay south if you want to go back to Foula to watch the wreck,' Simon suggested.

'Great,' I said. I looked forward to getting back to the island; I could feel myself pick up. Besides, a presence on the island would be a deterrent to other divers.

'When d' you think you'll go back?'

'As soon as possible,' I said, decisively. 'If they appear on the wreck, I'll let you know and we'll ask the court to take an injunction to stop them,' I added optimistically. 'I'll have some time to get the house repaired.'

Simon drove me to the ferry in Aberdeen. It was a bad night at sea, resulting in the ferry taking an additional twelve hours to make the trip. There was no point in trying to sleep, so I stayed in the bar all night, where a very agile Shetland fiddler kept everyone entertained. Prevented from crossing to Foula the next day, I bought a *Shetland Times* to see the headline: 'Salvage Men in Legal Battle'. I felt flattered that we were on the front page of 'the *Times*', as it was known, although in some people's eyes I am sure we were considered to be pirates.

I stayed at Walls while a storm 10 blew, but I got up the following day to find the wind had dropped completely, leaving an enormous swell with a glassy top to the waves. Against advice, I zipped across in the inflatable, tying myself into the boat, as the highest risk was being thrown out as it was tossed about on the largest waves. I was met by Rob, who helped me lift the boat out. As we stood on the slipway, I felt the darkness coming early.

The days had shortened and it was not helped by dense black rain clouds passing overhead. I had arrived there just in time.

For the first time on the island I did not care about the bad weather. Running across the track to the house, I pushed open the unlocked door just as the heavy rain started. Feeling my way along the dark, unlit flagstone passage, I passed the cupboard under the stairs and felt on my right for the door of the sitting room. Opening the door, the smell of burning peat wafted into the passage with warm air. I smiled. The faint glow from the fire showed that it was damped down. I loosened it up with the poker and it sprang alive. It was kind of Elizabeth to light it for my return. The two small windows let in little light from the dull afternoon, but the glow from the rekindled fire was sufficient to find the Tilley lamp. Fumbling with the pump and matches in the gloom, the vaporised paraffin soon hissed out to ignite on the mantle. A good strong white light reached out like a burst of sunshine.

I felt warm and happy as I looked around the room. Tins of diving-suit glue, brass plaques from the wreck, familiar books, diving suits slung over a chair and the red Roberts radio I had been given when my grandmother died all lent the house a familiar feel. These were the small trappings of my life. Compton Mackenzie, the author of *Whisky Galore!*, a man who loved small islands, was reputed to have said that on an island 'the individual is not overwhelmed by his own unimportance'. It may be true, as on the Scottish mainland I could walk through villages or cities never meeting anyone I knew, as if I didn't exist. On a small island everyone was known to each other, a character had to be earned, people knew each other too well for one to be assumed. I liked being on my own; it was not that Simon was difficult to live with, in fact quite the opposite – he was easy-going and little bothered him – but we were both thankful to get a break from each other.

When the steam started to rise from the kettle I prised off the round steel lid on the catering-sized tin of instant coffee, placed a measure in the mug, added the water and pierced a fresh tin of Carnation milk to top it up. Moving to the large, shabby mock-leather chair beside the fire, I considered switching the radio on when I felt a gust of wind. The back door must have been blown open. I cursed myself for not shutting it properly. Reluctantly I moved out of my chair, pausing when I heard the door close.

'Hello, hello?' I heard from the passage.

'Come in,' I shouted back, as I recognised the voice of John-Andrew. He was the youngest adult on the island. He worked a croft and helped where he could. Pleased that I was being visited, I hurriedly cleared the settee of the clothes I had dumped there. He opened the door, eyes flinching in the bright light of the Tilley; his dark beard and oilskins were dripping water on the floor. Undoing the top of his jacket, he exposed the large braces holding up his trousers – and unexpectedly, a bird.

'I shot a few scarfs today. Do you want one to eat?' He held out a large lifeless creature. I looked at it dubiously. 'I think in the south you call them shags,' he continued.

'Thanks,' I replied. 'How do you cook them?'

'There's nothing on the legs worth eating. I skin them and cut the breasts off. I'll do it, if you're wantin'.'

'That's kind of you' I answered, as I stood up and poked the fire, to get more heat under the kettle.

'Coffee?' I asked.

'Great,' he replied. He dumped the dead bird on an old newspaper on the table, giving his oilskin trousers a quick brush to get the water off, then slipped them below his knees and sat down.

'Sy coming back?' he queried.

'Not until the court case is over. I'm here to see if anyone from the mainland tries to work on the wreck.'

John-Andrew laughed. It was a pleasant laugh, then replied,

'The bad weather is in fer the winter, you'll never mak it out to the wreck.' Adding, 'How's the court case going?'

'It's been delayed for a week while the other side gets their argument together. If we lose, we'll have to leave the island. There's no other work for us.'

'To hell with them! Why not just work it anyway? You've taken the risk to find it.' His beard bristled with the emotion in his voice as he made up a cigarette from paper and loose tobacco, a common habit on the island.

'We'd be breaking the law and they'd put us in jail for contempt of court.'

'You deserve to win. The other salvage company thought the Shaalds were too dangerous to work.'

'If only life could be that simple,' I replied. 'We're in with a good chance, but there's no guarantee. It's only because our costs are so low that the *Oceanic* will be economic for us to work.'

'I'd like to learn to dive,' said John-Andrew.

Five years younger than me at only twenty years old, I had noticed that John-Andrew had a natural talent with his hands for both carpentry and mechanical work. He was also fearless and sure-footed on the cliffs, which would make him a good diver.

'Can you swim?' I asked.

'No, but I'm not wanting to stay on the surface,' he replied with his usual twist.

I was not surprised: most fishermen were unable to swim. With no intention of discouraging him, I replied, 'Diving is an odd business. It's just a method of getting to your work. The ability to complete the work's the important part.'

'It can't be that difficult.'

I looked at John-Andrew. Our eyes met. I thought he would find it easy to learn to dive, but was I qualified to teach him? He would be good, I had no doubt, because of his mechanical ability. If we won the court case, I decided, and if he was still interested and continued to show enthusiasm, then we would give him a chance.

'How do you know what to take off the wreck?' he asked.

'Apart from the big bits like the propellers, the bulk of the high value metals are around the engines. It's a matter of knowing how they work and how to dismantle them. It's quicker with explosives than using a spanner,' I replied.

Several days later it was pitch-black outside, with the wind blowing hard and more rain forecast, when I became aware of someone tapping on the window. I went to the back door to see who it was. Elizabeth stood there in a heavy coat and wellington boots. I asked her in, though knew she would decline. It was just a gesture.

'Simon phoned with a message for you,' she said.

'Any interesting news?' I asked with a smile.

Her eyes lit up. 'Naw, I'd not tell thee that, but the mail's been delayed from the Scottish mainland to Shetland by the bad weather. The court's delayed the case for two weeks.

'Mind and keep warm,' she said, as she turned away.

She disappeared up the road, only the odd reflection from her torch discernible as she dissolved into the darkness. Going back to my chair, I forced myself not to think of the court case, as it just made me cross and depressed. Instead I continued planning how we would work the wreck. There was enough money on it, but we had to make it a low-cost operation or we'd lose all our profits. It was a challenge, but I soon had it all worked out and relished the opportunity to get started.

The two weeks quickly passed: I washed clothes, dishtowels and bedding, soaking them in buckets of detergent before rinsing them in the burn, as our water tanks carried insufficient water. Lines of twine were put up in the sitting room and, with no more space, dishtowels and the remaining washing was draped on the backs of chairs in front of the fire. When this was completed, I developed a black-and-white camera film taken underwater on the wreck; it was well overdue. As there was no dark room, a black developing bag was used that had two cloth tubes to put my hands in. It was like a jersey with the bottom and neck zipped up. It was a fiddle to spool the film onto the reel, but easier doing it without Simon, as he always made jokes and tried to distract me. Once the film was on the spool, I slipped it into the light-proof plastic container, adding a developing chemical before rinsing it out with water and adding fixer. The final rinse required a trip to the burn, where I pinned one end of the film with two stones and allowed the water to run over it for twenty minutes to wash away every trace of the chemicals. Finally I hung it up in the sitting room using a clothes peg and allowed it to dry. Keeping myself busy was one way of getting the court case out of my mind: but it seemed to be lodged there, coming back to me every time I stopped.

I often looked out to the Shaalds, three miles from the Haa. It was now a maelstrom of white water, as enormous waves tumbled over each other, tripped by the shallow rock below. The water looked 100 feet high; it was the worst weather I had seen on the reef. I even wondered if it might be better to lose the case rather than suffer the possibility of humiliation of not being able to work the wreck commercially.

No, I was sure we could do it.

Working outside, I became fed up with digging drains and moved the remains of the pile of our recovered scrap at the top of the slipway to the Haa yard. After completing the job I went to the phone box to call the Receiver of Wreck in Lerwick, notifying him of its new position for his records. I was playing everything by the letter of the law.

'I've other bad news for you' he said.

'Oh no, what's it now?' I could feel my legs sag under me.

'There's two other parties enquiring about making a claim;

one's a commercial firm wanting to contact Nundy to offer to work the wreck for him.'

'And the other?'

'You'll know the other; it's a husband and wife, both divers. She's claiming she's been on the wreck. They've not spoken to me, but I hear the rumours.'

I knew her husband had been working professionally setting explosives underwater for an oil company. He'd had unexpected results – and was now unemployed. It was a timely and useful warning. Like anything of value, we expected bees round the honeypot, but I was surprised to hear a claim that someone had dived on the *Oceanic* before us. The islanders were unlikely to miss such an event. If she had not declared anything to the Receiver of Wreck, it should not be an issue, but if she joined forces with Nundy and claimed it was on his behalf, it could make it complicated. Living on my own, I let these rumours get out of perspective. My next call was to Simon.

A week later I was disturbed from my daydreams by the noise of a fishing boat coming into the pier. I walked down to see what was happening, only to see it backing off again. Another islander, Ken, was standing on the pier. Beside him stood a woman in her early forties. As Ken looked worse off for drink, I asked them both up to the Haa, Ken's home being a good mile away at the south end. The Haa was warm, the peat fire blazing away as they settled down by the hearth. I put the kettle on and went to get some milk from the kitchen. By the time I returned, Ken had his eyes closed.

'Is Ken asleep?' I asked

'I think so,' the woman replied.

'I'm Alec,' I said, introducing myself.

'Vanya,' she replied. 'I've known Ken for a while.'

From the unusual name, I knew she must be the person who claimed to have dived on the *Oceanic*.

'Are you a diver?' I asked.

'Yes, but I'm not going to dive on your wreck,' she quickly volunteered.

Uncertain of her motives, I asked, 'Have you dived on the *Oceanic?*' There was a pause as she thought what to say.

'If I were carrying out the salvage,' she replied, 'I would only take the works of art, paintings, sculptures and stained-glass windows.'

'Yes,' I replied. 'She was a unique ship.' A wave of relief swept over me: she must have thought the wreck was intact. It showed her inexperience of wrecks around Shetland and proved that she had never dived on the *Oceanic.* I was itching to phone Simon and let him know. 'Do you enjoy diving? Are you wanting to dive off Foula?' I asked, to move her off the subject of salvage and shift the conversation round to ask why she was on Foula.

'No, I've no intention of diving. I thought it would be fun to see the island.'

She was pleasant to chat to, but I was relieved when Ken woke up and they headed to his home. By then it was too late to phone Simon. I would tell him in the morning.

The fishing boat came two days later to take them back to the mainland.

Another week's delay by the court kept the tension up before Nundy's lawyer replied. He mentioned his employees had been sent up to Foula and been approached by us, asking for work. This was a break for us because he had been misinformed. I do not know who approached them, probably the people the Receiver mentioned, but the untruth, although unintended, would severely weaken his case. Now he would have difficulty questioning our competence, as his mistake made it obvious that he did not know who we were.

We made the most of his confusion in our reply to the court and then waited for Nundy's response. Simon kept me informed of the smallest information he could glean from the lawyers until our hopes were raised when Nundy allowed the injunction against us to expire. Finally he withdrew from the action. This was his way of accepting defeat.

'We've won!' said Simon. 'We're undisputed Salvors in Possession. We can work the *Oceanic* when the good weather comes.'

'Not as decisive as I'd hoped,' I replied, still unsure if this was cause for celebration.

'If he'd fought it further it'd have cost us a lot of money. He'll not come back now.'

We were given the opportunity to apply to the court for costs and damages in respect of our loss of earnings when we were restrained from working the wreck. We did not submit a claim and the dispute was ended with dignity, allowing both Nundy and us to get on with our work and put it all behind us. I was elated, but like all successes never quite sure the problems were over. It had seemed too easy.

I waited for a fine day to cross to the mainland, intending to start preparing new equipment for the good weather in the spring. Five days before Christmas I managed to make the crossing to Walls in the inflatable during a break in the weather and headed south on the ferry.

7

Oceanic Salvage

Leaving Simon in Fife, I set off for Shetland in February. Word had got round the salvage fraternity that we had found something exceptional and we thought one of us should be there to safeguard our interests. I hoped to get the mailboat to Foula during a spell of good weather. I was in luck. This was to be the first mailboat sailing since 24 December – a period of more than forty days because of the appalling weather they had experienced.

Leaving Walls through the Sound of Vaila, Jim pointed the bow towards the island while Rob, an islander who regularly acted as crew, covered some of the cargo with a tarpaulin. Twenty miles ahead of us, out in the Atlantic, Foula appeared like a vast lump of rock rising from the horizon; it is difficult to believe that it lies further north than Cape Farewell in Greenland. As we edged closer to the island I could see the land ramping up from low cliffs of between 80 and 120 feet high on the east side to the west side and the highest cliff, the Kame, at 1,220 feet. After three hours I could make out the entrance to the voe, with the war memorial clearly seen to the south of it. Excited at getting back, it was as if my eyes could not take in enough of the island. Arriving at the small exposed concrete pier, we unloaded the numerous boxes of food and several drums of fuel while Harry, the official postmaster, carried the mail bags up to the post office. The boat was lifted out by the crane and settled on its cradle before it was winched up to the top of the slipway well clear of any waves that might wash over the pier. Taking a shortcut to the Haa round the side of the winch shed and scrambling up the bank onto the rough track, I stopped and looked up. Smoke was emerging from two chimneys and giving the familiar smell of burning peat. I was back and could see Elizabeth standing at the door.

'It's aye a bit damp. If I'd known you were coming earlier I could have dried it out a bit,' she said.

'It's just nice to be back,' I replied.

Elizabeth beamed one of her smiles. 'I never knew if you'd ever be back with yon court case,' she said, adding, 'Come up to the Ham for a cup of coffee when you're settled in.'

'Are you coming in?' I asked.

'Naw, naw, it was the least I could do to light thy fire,' she replied before turning to leave.

Simon had regularly kept in touch with Elizabeth. We had sent instructions to our Shetland lawyer to try and obtain something positive from the court to enable us to make an agreement with the wreck owners, who were still concerned, as Nundy had only withdrawn; if the court had made a ruling in our favour against him, it would have been different. We required the owners' agreement to obtain an explosive permit; without explosives, our recoveries would be limited. Simon was also organising a cook for part of the summer, as it would increase our working time. Neither of us enjoyed cooking – we would come in from the wreck usually cold and starving, desperate for food, having been unable to eat much because of the diving, then have to wait half an hour or more as we prepared warm food. When we were busy, this chore usually fell to Simon, as I was more suited to the mechanical side of the business, preparing the gear for the next trip. It was a division we were both happy with. On Fair Isle, Deidre, a student, had been a summer cook at the observatory and she had often visited our hut and cooked us the occasional meal. When we first went to Foula, she came across to see the birds and liked the island, so Simon called her during the winter and she looked forward to coming over for a few months during her summer holidays. She would cook one meal a day for five days a week; in return, we would pay her a small wage, plus free board and lodgings. We knew Deidre and did not anticipate any problems. She had only one fault we knew of: she continually beat us at Monopoly.

The islanders considered April the earliest time the weather would let us return to the Shaalds, but we were both raring to go and had no intention of missing any working days. By the beginning of March I was rushing to complete jobs on the house. I had rebuilt the fire, repaired nearly all the drains, replaced missing slates and started to work on the wood in the ceiling of the leaking front porch. This was where we hoped to store some of our equipment. The sand and gravel for the drains had to be barrowed up from the beach, a laborious, time-consuming task that usually involved old Tom latching himself on to me. He would take his unlit pipe out from between his worn, blackened teeth and say, 'Man, I'll tell thee . . .' before launching into a repeat of the shipping forecast as if it was his prediction of the weather, or reciting a bit of news he had heard on the radio. I never wanted to be rude but it became difficult to get away from him, until I was helped by a warning issued on the radio stating some poisonous drums had been washed off a ship near Foula. As soon as I came close to our house and the sea he turned back.

I enjoyed time on my own in the evenings, sitting in the warmth of the peat fire and reading a book, doing sketches of the wreck or working out the design of equipment for a boat I hoped we could buy. At times I just sat there and dreamt, thinking of the fantastic opportunity ahead of us, and where it might lead. I realised how many of my friends were getting married or 'going steady' and how my parents had always instilled in me the concept that I should not get married until I could provide a home and regular income. Their marriage would be deemed old-fashioned; my mother had given up work after she was married, whereas most of the girls I liked had jobs, and one in particular who I had been going out with had moved south to improve her prospects. I was fixed in this work, there was no way I wanted to be anywhere else, but it seemed an impossible situation in which to keep a relationship going, working a wreck on Foula. I doubted any women would want to give up their good job and come and live on Foula in a house without electricity or water, however romantic it might sound.

Few girls passed through, anyway – birdwatching at that time was mainly the interest of men, but those girls that did reach the island I usually liked because they were often strong characters. In Fife I was aware of some friends who fell head over heels for someone; it was as if a door to their soul had been opened, as I watched their priorities and ambitions quickly change. I wondered if it would happen to me.

When Simon joined me on Foula, he brought an enormous load of equipment with him on the ferry north, which was added to at Walls with the gear stored on our Halflinger; this had come from Fair Isle but we never managed to get it onto the island. Kenny, the owner of the small fishing boat the *Lustre*, which we had previously used, was now working on an Orkney trawler and he told us to use his boat. We worked into the night to load the Halflinger, our equipment and some sheep feed Elizabeth had ordered onto the boat. After a few hours' sleep we set off for Foula at seven, Simon putting his red, white and black striped bobble hat on as we left the pier. It was like a symbol that we were about to start our island life again. Arriving at 9.30 a.m. on Foula, Jim helped to unload, giving us a quick turnround to be back in Walls by 1 p.m. After tying up the trawler, we returned to Foula in the inflatable.

Two days later we had our first dive of the year on the *Oceanic*. It was fantastic to see this enormous wreck again and, after a good search to reassure ourselves of its value, we secured new moorings and recovered some copper and brass. The tide was as bad as we had experienced, the slack water lasting barely twenty minutes before it gradually turned into a river. Jim came out to meet us when we could no longer work on the wreck and we returned to the island to search for the *Nordstjernen*, a motorised schooner that was lost on the east side of the island. Jim gave us a position and Simon slipped into the water while I kept the inflatable above him. Looking over the side we could see Simon on the bottom 30 feet below; a section of shaft with a propeller on it was clearly visible.

'That looks like the propeller,' said Jim.

'You've seen it before, Simon.' I smiled as we watched him swim towards it.

The other items to look for were the ship's anchors, which Simon found. We quickly lifted the *Nordstjernen*'s propeller using a lifting bag and towed it back to the voe, where Jim placed it on the pier with the crane.

We discussed the size of the anchors, as, thinking ahead, when we bought a salvage boat and brought it to Foula, we would have to lay a mooring in the entrance of the voe behind a reef called the Head of the Baa. The mooring would provide little shelter from any wind with 'east' in it and would therefore need good anchors. The pier was not only too shallow for a big boat at low water, but it would be impossible to lie safely in any swell.

Beginning to get back into our work routine of charging the

air bottles, changing in and out of diving suits, refuelling the boat and shifting our recoveries off the pier, I felt this would be a great summer. Listening to the early morning shipping forecast at 5.50 a.m. allowed us to plan the day. If the tide times were favourable, we would be on the wreck site by first light. Sometimes, if the forecast indicated bad weather and it matched what we saw out of the window, we could go back to bed, but if it still looked good outside we often managed to dive before the bad weather arrived – on occasions, however, we'd find it freshening as we went out to the Shaalds and it forced us to return. Now using two inflatables, one towed behind the other like a truck and a trailer, we secured the first inflatable to the mooring rope, pulling the second one alongside to be filled after the first was loaded. This, along with our increasing knowledge of the wreck, nearly doubled the scrap we could lift on the end of a rope and transport back to the pier.

Simon usually dived first. When his air bottles were empty, I would go down and take my turn. We were able to empty the air bottles underwater because the wreck was so shallow; we knew when we were running out of air, as we literally had to suck the air out of the bottle. Our pressure gauges had been removed, as they had a habit of catching on the wreckage. We only stopped diving when the tide was too strong for us to work: depending on the weather and the moon, this slack-water period lasted from as little as twenty minutes up to a maximum of two hours. We don't believe we experienced the 12-knot speed stated in Professor Holbourn's book, but at times when we could not dive it was probably running at nine knots. With the outboard started, Simon would 'let go' the mooring and the two inflatables would swing round, one tied behind the other, and drift back on the current before the outboard was put into gear. As we motored back, an eye was cast over our recoveries, often finding neatly fashioned brass edging and fittings that reflected the high quality of the ship.

Thomas Ismay, who formed the Oceanic Steam Navigation Company in 1869, later to become the White Star Line, launched the first *Oceanic* in 1871. He was one of the first ship owners to place the first-class accommodation amidships, where there was least engine vibration. He included all classes in his improvements and sent a Captain Hinds on the first *Oceanic* as a steerage passenger to report on his experience at this lowest level of accommodation. As a result, Ismay had rooms built for married couples and families in third class. Previously they had stayed in a large room holding twenty to thirty people, having to provide their own beds, blankets and eating utensils. His improvements were revolutionary in ships carrying emigrants to America and the second *Oceanic*, the one we were dismantling, incorporated these improvements.

I always felt refreshed after an early morning dive and my stomach was crying out for something to eat. Once back, we would throw the scrap and diving gear onto the pier and walk up to the Haa, where we took our near-watertight wetsuits off outside the back door. Struggling with the tight fit, I would bend over, resting my head on Simon's chest as he gripped the wetsuit top at either side of my waist and pulled. This was always hard: when it was halfway over my head it felt like I was in a straitjacket; I would be trapped and unable to move. Simon and I had agreed we would never play tricks on each other while removing a diving suit. A final pull and it was off, and I could breathe and move again. I would then do the same for him. The bottom section unzipped from the neck down and looked like full-length long johns. They came off like tight-fitting trousers. We would throw the wetsuits on the wall and run naked into the Haa, grab a towel from above the fire and dress as quickly as possible.

Clothed and warm, I would return to the inflatable, charge the air bottles and refuel the outboard while Simon knocked up some food. It was dangerous to dive on a full stomach because of stomach cramp, so we had to eat sparingly until the day's diving was completed. If we had bread we would cut slices off whole-sale-sized blocks of cheese that were about 14 inches long and four inches thick or use jam and marmalade from catering tins. In the evening we'd treat ourselves to a big meal, but if exhausted we'd put in little effort, opening tins such as a Fray Bentos steak and kidney pie, corned beef, Spam, Heinz baked beans, sausages in beans, or Campbell's meatballs. Our vegetables consisted of powdered potatoes, tins of carrots and peas and freeze-dried veg. Pudding might be Ambrosia creamed rice with tinned fruit, tinned custard or custard powder made up with powdered milk. Occasionally as a treat we would make up Angel Delight and open a tin of Carnation milk for our coffee or tea, otherwise it was cartons of long-life milk or, in extreme circumstances when these had run out, powdered milk. As well as learning about explosives with Chris, I had discovered which tinned foods were worth eating.

With no diving, we had plenty time to cook. Simon could make spaghetti Bolognese, or 'spag bog', as he called it, and cheese omelettes always provided a quick meal. Food on the island was a much-valued commodity, as you were never sure when supplies could be replenished, so very little food was wasted. Rather than dump the mouldy bread, I would scrape off the mould and make bread and butter pudding. If the bananas were overripe, as they usually were by the time they had reached their final destination on Foula, I would make banana bread.

The islanders made soda bread and had no reliance on imported bread, although Maggie, the nurse, was the exception and made bread with yeast. She gave me lessons in baking until I had grasped the basics. At the beginning I was unable to get the dough to rise evenly due to the drafts, but she suggested placing the bowl of dough in a plastic bucket with a towel over it and leaving it near the peat fire. It worked; the freshly baked bread liberally coated with butter, jam or marmalade rarely lasted more than an hour and disappeared even quicker if we had visitors. We obtained slabs of live yeast from the baker in Scalloway and on the occasions I ran out a fishing boat would drop it off at the pier.

After we signed a deal with Andy Beattie, the managing director of Hay & Company, who owned the *Oceanic*, we went to the police, who issued an explosive permit. Picking the explosives up from a decrepit store outside Lerwick, we loaded them on the inflatable and made the trip back to Foula. Although the islanders were completely honest – as Harry said, 'You could leave a £5 note under a stone on the pier and it'd be there until it was blown away or rotted' – we required a secure store for the explosives to comply with the law. With limited options we decided to keep the boxes under the stairs and the detonators in the bottom drawer of the chest of drawers in my bedroom.

Simon took little interest in the explosive work, which was down to me. I enjoyed setting the charges, doing a rough sketch of the laying of the charge and the quantity needed. In the Firth of Forth I had used large quantities of explosives but was now able to undertake precision work using very small charges, thanks to my experience with Chris and our summer on Fair Isle. In the evening we prepared the explosives for the next day, usually making up multiple charges, up to forty at one time, and these were linked by Cordtex, the white explosive cord that looked like washing-line cord I'd first seen Chris use on the *Hurlford*. It burnt at 20,000 feet a second, which can be considered as instantaneous. Gunpowder burns at around 1,000 feet a second. For use under water we cut the Cordtex into appropriate lengths and sealed the ends, tamping the explosives in the cut end by using a piece of wood the size of a match and then filling the half-inch space with glue to prevent the water penetrating the cable. We ran a 'main line' of perhaps 100 feet with over forty 'branch lines' taped onto the 'main line'. Each branch line ended with a small stick of explosives that could be added to underwater if a large charge was required. The Cordtex was initiated by a Hydrostar underwater electric detonator that was securely taped at the start of the main line. We rarely had failures and were quick at blowing off the many pieces of non-ferrous on the wreck. Setting the explosives was always the last task before leaving the wreck at the end of a tide and we were careful to look out for old explosives that had failed to go off or any form of munitions that were originally aboard the ship.

When the *Oceanic* was built, the Admiralty paid the White Star Line to reinforce the decks to make them strong enough to take mountings for 4.7-inch guns in the event of a war. At the start of the First World War, when she was in Southampton, the guns were quickly fitted and she sailed out as RMS *Oceanic* – an armed merchant cruiser.

Alongside one of the propeller shafts on the wreck was a pile of the 4.7-inch shell casings. We had come across cordite before and I had bought a 1905 *Handbook of Ammunition* originally supplied to the torpedo office on HMS *Vincent*. It was beautifully illustrated, with coloured drawings of the various fuses, shell cases and other munitions. This along with some advice from fellow salvors gave us the confidence to remove the cordite from the shell cases underwater; any we couldn't remove was piled up next to an explosive charge. The blast burst the brass shell casings but the spaghetti-like filaments of cordite within the casings did not go off. We remained super cautious, reminded of our time on Fair Isle when we heard the blast of a torpedo going off on the deck of a fishing boat 15 miles off the island. They had caught it in their nets with fatal consequences. After blasting we used the following tide to clear up all the smaller pieces and try to estimate the weight of the larger bits.

As large pieces accumulated Simon agreed to buying a salvage boat. There were also five spare single-propeller blades weighing six tons each that could be lifted clear of the wreckage with little, if any, blasting required. The main propellers were huge; they were only six inches less in diameter than the *Titanic*'s, with more than 29 tons of brass in each complete propeller. Because they were partly buried beneath rock and wreckage they would require considerable work to be able to place the explosive charges to blast them off the shafts. It was something we would do later, as there was plenty lying there to lift.

Everyone was welcome at the Haa. We enjoyed the company – a few strangers had already stayed with us, coming across either to see the island or the birds, when their tent had suffered a mishap resulting in them becoming unbearably wet with no way to get dry. Our spare room had two beds, and with a warm fire in the sitting room and plenty floor space we could accommodate a few. Our unexpected visitors rarely stayed long, taking the first opportunity to leave the island on the mailboat or a passing fishing boat, in case the weather broke and they were stranded for weeks or months.

In preparation for Deidre's arrival we tidied up the Haa as best as we could, removing old newspapers, diving suits and clothes from the floor before sweeping it. The kitchen was scrubbed, including the stone floor. A grey Aladdin paraffin heater given to us by the departing school teacher had its wick trimmed and cleaned before it was placed in her bedroom. It was not perfect, but a noticeable improvement – to us. Although Simon had agreed basic terms with Deidre, it was all flexible and we hoped it would develop into a system that suited everyone and give her plenty time to go out birdwatching, which she loved.

The appointed day came and she phoned Elizabeth to let us know of her arrival at Walls. I set out in the inflatable from Foula as the mist came down. It was a foolish thing to do. I soon became surrounded by it, unable to see 100 yards in any direction. It was eerie and the compass was inconsistent in the bumping inflatable, the needle moving backwards and forwards with the motion. Unsure of the compass, I stopped at times to let it settle. It was the first time I had been out with the inflatable in thick mist and felt my confidence dwindling. I knew there was no return because Foula would be harder to find than Shetland, and that (I hoped) I couldn't miss.

When I had overrun my journey time by more than half an hour, I became worried. Eventually I made a landing on a stony beach with no idea of where I was. Walking inland, I found a Shetlander and asked him. He told me I was a long way from Walls, but he was kind enough to take the inflatable on his pick-up and drop me there. After meeting Deidre, I waited for the mist to clear and the following day we made the return journey. It was a timely warning: I had become quite casual about making long trips in the inflatable, and I needed to be more careful.

On Deidre's arrival our work output improved and meals became something to look forward to. The only downside, as we had predicted, was that she continually beat us at Monopoly.

Simon started to make arrangements for selling the recovered scrap. He intended to charter the Fair Isle mailboat again. I scoured the *Fishing News* and other commercial boat magazines to see what boats were available at a suitable price for us to buy as our own salvage ship.

'Are you sure we can't continue with the inflatables, perhaps hire a boat occasionally?' asked Simon, hinting at his concern.

'We'll never recover all the big bits,' I replied.

'What size of boat are you thinking?'

'We want one we can anchor in the voe,' I said. 'It limits its size but we'll not get much work done if the boat can't work from Foula. If we operate a bigger boat from Walls or Scalloway, our costs'll rocket and we'd lose a lot of time.'

'Small sounds OK to me,' admitted Simon.

'It's a good time to go south. The May Water will be here soon.' We'd been told this occurred when the plankton grew at the end of April and the beginning of May. It spread through the water in clouds, making you feel that you were diving in a near-zero-visibility soup for two weeks. It reminded me of the Firth of Forth and I didn't fancy it.

When the *Good Shepherd* left Fair Isle, we listened for her on a fishing boat's radio. They said they would give a call to let us know when they would arrive.

'*Oceanic, Oceanic, Oceanic*. We're expecting to be at the Foula pier about seven.'

'Must be the first call to the *Oceanic* in sixty years,' said Simon.

'Well, boys,' said Elizabeth when we told her, 'I'll let some of the island know.'

The *Good Shepherd* was loaded within an hour because of the help given by the Foula men, who all turned up, old and young. We had a calm trip to Wick on the Scottish mainland, arriving at 10.30 a.m. the following morning, then immediately discharging the scrap onto a waiting lorry. All went like clockwork until the customs men arrived.

Simon and I had all the correct paperwork from the owners of the wreck and the Shetland Receiver of Wreck, but they then tried to charge the *Good Shepherd* light dues. This was a payment for the lighthouses and navigation buoys, paid by cargo vessels; fishing vessels paid at a lower rate. On this occasion they treated the *Good Shepherd* as a cargo boat, with correspondingly high charges, because she had been carrying our scrap. The islanders resented paying additional charges for anything they did not use; it was hard enough running the mailboat without any additional costs. We hoped they'd put a good case forward for not paying. However, we needn't have worried. While we were leaving with the loaded lorry, we heard the Fair Islanders asking if 'the charge was for daylight or moonlight'.

The sale of scrap in Glasgow gave us our first large cheque as a partnership, and we were able to open a joint bank account entitled Crawford and Martin.

With the money in our pockets, we set off for Peterhead, where we found *Trygg*, a fishing boat that had been used as a trawler before her present job of running supplies and equipment out to oil-related ships at anchor in Peterhead harbour bay. She had recently been replaced with a more modern vessel but was still used on occasions when the other boat was busy. The conversion from a fishing boat to her present job had included the removal of most of her accommodation to extend the hold. This had made her almost unsaleable as a fishing boat, resulting in a bargain price. Built in Norway in 1938, at 49.9 feet long and 17.4 feet wide, she had a relatively new 120 HP Kelvin engine and a double-barrelled Fifer trawl winch. Her pine wheelhouse was attractively wood-lined and contained a compass and VHF radio. The hull was painted black, the wheelhouse white with a varnished pine front, giving her a tidy look. Her steel mast and derrick had been used for transferring the stores and would have to be beefed up a bit, but otherwise she was ready to go. Lifebuoys, a life raft and numerous life jackets were aboard to comply with her present use where she might be asked to change ships' crews as well as stores. Best of all, Simon was happy with her and we quickly came to an agreement we should buy her. We haggled over the price, although the already-low asking price reduced the financial risk if we discovered we were unable to operate her from Foula. After concluding a deal we immediately hired a welding plant to lengthen and stiffen up the steel mast, using a new, larger-diameter tube that was purchased from a local scrap merchant. After three days' work that involved working late nights, we set sail for Foula exhausted.

On tying up at Foula we finished the rigging work on the

extended mast and derrick and then took some of the young ones on a tour around the island: the natural arches, rock faults and cliffs on the north and west sides made it a stunning trip; the whole place was alive with seabirds. That night the wind changed direction. Fortunately we were staying on the boat and realised when the motion started and the pier become awash. We quickly left for the shelter of Scalloway. The islanders had always warned us that it would be difficult, if not impossible, to keep a boat at Foula, and a fisherman had described the weather on Foula as 'eight months of winter followed by four months of bad weather'. Now discovering it might be true, but not wanting to waste time, we used the days of bad weather to fix up a heavy lift point at *Trygg*'s bow for lifting the *Oceanic*'s propellers. We asked the boatyard in Scalloway to help us with the strengthening and worked with Jack Moore, the owner of the boatyard, in fitting a large iroko beam just forward of *Trygg*'s mast. It was a time-consuming job and Jack used his forge to make a special ringbolt that went through the 18-inch square wooden beam and would take a load in excess of 15 tons. It was late Saturday evening before we stopped working.

'Well, Alec, I'm stopping for the night,' Jack said. 'I don't work on a Sunday, but here's the key to the store. If you take anything, just mark it down in the book.'

'Thanks,' I replied. 'I'm sorry you've had to spend your Saturday evening with us.'

'I've enjoyed it. I'll see you Monday morning,' he said before leaving.

Simon and I looked at each other. 'We only met him this morning and he's trusted us with the keys to the store,' I said.

'It's Shetland,' replied Simon, and I knew what he meant. Jack had worked all day with me. I was surprised he hadn't delegated the work to one of the many men he employed, but I was pleased he hadn't, as I enjoyed chatting with him while we worked and he took a real interest in our salvage, asking many technical questions.

'When d' you think you'll finish the job?' queried Simon.

'We'll be ready to sail on Monday morning if we work all tomorrow. We can rig the large pulley blocks we bought at the Fife shipbreaker's on Foula.'

'Good,' said Simon. 'The shopping's done, but I expect there'll be a few boxes delivered before Monday.'

Now that we had a proper boat, we had been given long shopping lists. Apart from fuel, most of the items were small, made up of things that could not be bought over the phone at the Walls shop, requiring the more specialised shops in Scalloway and Lerwick.

We returned to Foula on Monday with a load of passengers for the island; it was to be something that became a regular feature of our trips, as knowledge of our new boat spread.

8

Spoils Gained but Nearly Lost

Lifting the propellers would be the most dangerous work we thought we'd carry out; from experience I knew accidents occurred at the most unexpected times. We rigged the heavy lifting blocks and took *Trygg* out to the wreck for our first lift on a six-ton spare propeller blade, thinking it would be the easiest. The blocks allowed the three-ton trawl winch to have a lift of 15 tons on the seabed, though the penalty for this reduction was the lift taking five times as long to haul in. The slower speed would only affect us as the blade came off the seabed. If there was a heavy swell, it would bounce on the seabed until it was clear.

Deidre came with us to help me on deck while Simon dived. The three-grooved sheaves on each block had the wire rope around them similar to those on a big crane; these hung over the bow. Deidre lowered the winch as I fed the wires over the bow, letting the bottom block disappear below the water. Simon attached a rope to the propeller blade and we pulled the bottom block close enough to the propeller blade for Simon to secure it with a sling. Because the heavy blocks were attached at the bow I had no concerns about the boat capsizing. I did not want a repeat of our experience with the *Lustre*, but it was important that the lift was completed within the guaranteed half-hour slack water – after that it would become difficult, if not impossible, and we would probably have to cut the wire to release the boat. As the action started I became like a football supporter on deck, urging the unhearing Simon on as he worked underwater.

'Come on, come on, Sy,' I would be saying, urging him to get the sling attached as quickly as possible, although the only person who could hear me was Deidre, who gave me odd looks.

We both watched his bubbles. I was continually glancing at my watch and at the water to see if the current was gaining in strength.

'What is he doing? What is the problem?' I repeated again and again, knowing he would have the hundredweight-and-a-half sheave block swinging backwards and forwards above his head as he tried to connect it. It was a dangerous job, but as a spectator there was nothing I could do but worry. Deidre, totally relaxed, just looked at me, wondering why I was so agitated, unaware of the dangers Simon was facing.

When Simon eventually came to the surface to tell us to haul in on the winch, I went aft to operate the levers while Deidre remained at the side of *Trygg* to relay messages from Simon.

As the winch hauled in, *Trygg* was pulled directly over the blade, a heavy load coming on as she started to take the weight. The bow sank deeper in the water. I stopped the winch to go forward to check the iroko beam we fitted, but it was not showing any sign of movement. Deidre was leaning over the side talking to Simon, who said there was little more he could do. We helped him onto *Trygg* to keep him clear of the underwater lifting gear in case something broke.

It took a few minutes for the blade to break free and shake off the wreckage that lay on top of it. The bow of *Trygg* was well down in the water and her stern raised at an odd angle, the blade bumping on the wreckage below us, sending shock waves through the boat until we had it lifted clear. Releasing the mooring, we steamed slowly towards the Foula pier with the six-ton lump of brass suspended below our bow. It was a tense and slow process. Concerned at the head-down angle, we checked forward to find water trickling into the boat through the overloaded bow. On the outside of the bow the paint was visibly cracked between the planks and the caulking (sealing material) was being squeezed out. But there was no sign to indicate a sudden failure in any of the woodwork. We continued with the blade swinging below us in the swell, the top block rubbing against wooden protection we had added to the bow. I hoped it wouldn't rub through it.

As we closed in on Foula the motion became less, but I realised we were becoming more bow-down than we had been on the reef. Entering the forward compartment to investigate, I saw the water was no longer trickling but pouring in. The pump suctions were all aft, making us unable to pump out the boat with either the engine or the emergency hand pump. Approaching the pier I raced in with the inflatable to bring some of the watching islanders out to help run a chain of buckets to keep the water down while I extended the engine-driven pump intake further forward.

It was high water when we came alongside the pier, the extra depth allowing us to drag the blade close to the mailboat crane until *Trygg* grounded. The blocks were released and Trygg backed out before she was caught by the falling tide, which would trap her on the rocky bottom. Simon dived in the shallow water to hook the blade to the crane. Everyone watched as Jim operated the crane, the blade rising slowly out of the water before being swung round and placed on rocks. Simon and I looked at each other – it was the vindication of our purchase of *Trygg*.

It was time for a brief celebration for everyone in the Haa before Simon and I made some improvements to the lifting gear, which would make the next lift easier. A few days later the weather was calm enough to set off again to try another lift. We lifted four blades in one day, each during the slack water between tides – a total of 24 tons of brass for our day's work. Only another six blades, then it would leave us the twelve-and-a-half-ton central bosses to be lifted and that would be the propellers dealt with. We still had to blast these bosses and some blades free, but they would have to wait.

Meanwhile we took advantage of the good weather by lifting large objects under *Trygg*'s bow that were too heavy to lift on deck. We dropped them in 20 feet of water in the partly sheltered area of the voe where they could easily be broken up and lifted aboard when the weather was too bad to work on the wreck. Knowing there was some bad weather coming and we might have to leave Foula at short notice, we lifted a propeller blade into the hold of *Trygg* using the mailboat crane at high water. She would only take one blade before she was sitting on the bottom and we had to winch her off.

During the night in pitch-black driving rain, we were forced to leave Foula, the wind increasing on the way across to gale force. During the crossing, she received such a battering that the damage at the bow caused her to leak badly, the water pouring in, the pump struggling to cope. As we approached the islands guarding the entrance to Scalloway, I was not confident enough to be sure where the reefs lay close to the channel and reluctantly dodged on the seaward side of the islands with all the pumps running until there was enough light to guarantee a safe passage. I regretted not having radar.

Arriving in Scalloway, we had no choice but to haul her up on Jack Moore's slipway. Fortunately, there was little physical damage apart from rubbing, but much of the old weather-hardened caulking at the bow had come out and had to be replaced, along with some bolts securing the stem. It took two days to repair, along with the separate work involved fitting a second-hand radar unit.

We rarely left Scalloway with an empty hold – on this occasion a tractor, trailer, digger and two goats arrived on the pier to be taken to the island.

'You're not taking all these,' exclaimed Robbie, who was re-fuelling us.

'Yes,' I replied, as I too wondered how we would fit them all on.

After we had loaded all the equipment, we heard that a government ships' surveyor was due in Scalloway the next day. As *Trygg* had no certification, we should not have been carrying cargo. Normally the cargo lay unseen in the hold, like the fuel drums we carried or scrap, but in this case the tractor and other agricultural equipment was on deck, as it wouldn't fit through the hatch to go below. We decided to leave that evening in an increasingly heavy swell. It was not a wise decision, and soon water was pouring over the deck, washing the goats off their legs, even though they were in the shelter at the back of the wheelhouse. Two people had to stay on deck to look after them.

I was reluctant to turn back, as we had an improving forecast and there was a risk the deck cargo might shift when we turned, causing her to capsize when she was broadside to the waves, so we ploughed on, hoping the weather might improve. Not until we were in the shelter of Foula could I relax. Leaving the boat to unload in the morning, I was relieved to get to my bed in the Haa.

With the weather unsettled, we spent less time on Foula, as we had to run for shelter at Scalloway. On each trip we had passengers going to or coming from the island. The second-hand radar unit made an enormous difference to our confidence in bad weather and fog, but the islanders had been right – it was almost impossible to work with *Trygg* from the exposed pier on Foula. To solve this problem, we lifted one of the *Oceanic*'s seven-ton anchors, complete with a length of chain, placing it on the easterly, most-exposed direction at the entrance to the voe, and added two anchors from the wrecked schooner on the north and west side of the voe. John-Andrew had now joined our team and helped us with the mooring; it was a good place for him to get some practical diving experience, although he had taken the initiative by making his own diving suit and dived in the voe to recover condenser tubes and small bits of brass that we had accidentally dropped over the side. He proudly gave us his recoveries – how could we not give him work?

We began to shift the vast stock of brass lying in the entrance to the voe. Another propeller blade was loaded into the hold of *Trygg* for the trip south to Fife; it was so tall it stuck out of the top of the hatch. The hold was filled up with smaller scrap, and hatch boards were cut and canvas nailed around the tip of the blade to keep the hold watertight. By the time she was full, *Trygg* was well down in the water, with little freeboard (the height remaining between the waterline and the deck).

The forecast was a maximum of Force 6 for the sea areas of Fair Isle, Cromarty and Forth. It was as good as we could expect. By the time we had steamed six hours, the wind had freshened, causing a heavy sea. As darkness fell it had become a full-blown gale, but *Trygg*, with her heavy load, was steady in the water, giving us no concern. I regularly checked the engine, followed by the pump in the hold, but it was as dry as a bone. Taking three hours each at the wheel allowed the other to sleep, make cups of tea, eat or read a book.

Simon had completed his stint, so I took over from him. The wind was now head-on, but we were making good progress. I checked our position with the radar to discover we were exactly where we were meant to be. The pitch-black darkness outside was only broken by the reflection of the navigation lights in the heavy spray of water that was breaking over us. The engine sounded well and the boat was much more comfortable with the heavy load, even though the deck was awash all of the time. I knew there was no threat to the canvas cover on the hatch; it was well wedged in with nailed planks of wood to be absolutely sure, and I had checked it again before taking over from Simon.

So the next thing that happened took me by surprise. There was no warning at all. I saw it loom over the bow, lit up by the mast-head lamp. I closed the throttle as an enormous wave rolled over us. The boat stopped as if it had hit a brick wall.

The door to the compartment behind the wheelhouse flew open, as Simon was propelled from sleeping on a bench bunk in the after end of the wheelhouse to lying on the floor beside me. I could see the water on the deck – it was up to the bottom of the wheelhouse windows. I stood stock still. Everything seemed to overwhelm my thoughts. I was not sure what to do until there was a sudden crash from the engine-room hatch at the back of the wheelhouse, followed by the sound of water pouring down below.

The engine stopped.

'What's happened?' shouted Simon above the noise of the water.

'The engine room's flooding,' I said.

Simon picked himself off the floor, dazed, and looked out the wheelhouse windows in surprise, trying to take in the situation. The mast at the bow stuck out of the water as if there was no deck below it. I felt the boat wavering with the weight of water on her. There was a pause in the waves, and I looked to see if there was another big wave coming. I could see white water washing over the deck; the next wave must have broken, giving us a moment of calm.

Trygg had to make her decision: would she sink or float?

I felt her slowly rise, the foredeck appearing as she began shedding the water from her decks, shaking it off like a dog.

'What can I do?' asked Simon, as he stood beside me with relief.

'Can you use the hand pump on deck? I've got to get the engine running.' Without hesitation, he threw on some oilskins and went out of the door, hanging onto anything he could find as he worked his way round to the front of the wheelhouse to the emergency hand pump. The sea looked vicious outside, the water washing up to his waist as it tried to pull him over the side. I grabbed a small mattress from the bunk and some bedding before entering the engine-room hatch in the after end of the wheelhouse. Standing on the access ladder, I jammed the canvas and bedding against the broken hatch that led to the deck. It slowed the water entering through the splintered wood. Satisfied it would hold for the moment, I slid down the rest of the steps, my feet disappearing into the oily water that was sloshing from one side to the other as the boat wallowed. She must be lying sideways to the wind, I thought.

I held onto the ladder, worried that I would be washed off my feet. Carefully placing my submerged feet on the floor plates at the side of the engine, I felt the cold water rise above my knees. Oil cans that had been in a rack at the side had been torn loose and were banging against the engine like giant bath toys, oil pouring out as she rolled. Above the noise of the banging I could hear the waves breaking above me on her deck. I hoped Simon was OK. The water swilling around gave off the pungent smell of the bilges. I tried to secure the batteries in their shattered box behind the engine. The fumes of the spilt acid had mixed with the salt water to form a gas that burnt my throat and nose as I bent down to replace them with the two undamaged spares.

It must have taken a quarter of an hour before I could see the water level dropping from Simon's pumping, but it was still too high for the engine to run. I caught the floating cans and debris, throwing most of it behind the fuel tanks, where it became trapped. I knew I couldn't wait much longer before trying to start the engine: the rolling was getting more extreme, and we would be much safer if we could turn her and run before the wind. Would the starter work? As the water level dropped below it, I took the cover off the back to drain the water and wipe it out with a rag before protecting it from splashes with a piece of canvas. I slackened the engine's air intake manifold to see water drain out of it – that was why she'd stopped. Depending on the damage, she might not run again. Once the water level had dropped a few more inches I could try and start her. I hung on, unable to keep my feet at times, with the extreme roll of the boat.

Climbing up the ladder to the wheelhouse, I partially opened a wheelhouse window to see Simon still pumping. He stopped to hang on when a large wave swept across the deck. I wondered if I should call someone on the VHF radio now the batteries were back working, but it only had a range of about 28 miles and we were over 40 miles from land. There might have been a fishing boat near us – but no one could do anything in the next few minutes other than have difficulty taking us off if they were nearby. I wasn't going to waste time; our best option was to get the engine going.

'OK, Sy!' I shouted out of the window.

He saw me and gave a thumbs-up between strokes of the long-handled Whale pump. We both knew we were in survival mode, with no thought of anything other than saving the boat and ourselves. The two seemed inextricably linked.

Returning to the engine room, slightly fearful of the extreme rolling the boat was doing and worried that the scrap might shift, which would take us quickly to the bottom, I threw over the decompression levers for each cylinder to let the engine blow the water out of the cylinders when it turned. I pushed the starter button, willing the engine to turn over. The light in the engine room dimmed; the starter sounded laboured. The battery power must be low, I figured, but it was working, I could hear the engine turning, the large flywheel picking up water and throwing it up in a shower. I held my breath until she had turned a couple of revolutions before throwing over one of the decompression levers. I heard a cylinder fire, then threw over the other levers until it was running. It sounded rough. I checked the oil pressure and engaged the mechanical bilge pump. If I could just get the engine room dry I could have a better look at the engine. After a few minutes Simon's head appeared at the hatch, looking down at me.

'I see you've got the pump discharging water.'

'I don't know how long she'll run. Can you put her in gear and turn her head round to go with the wind?' I shouted above the noise.

'What about water in the hold?' he asked.

'I'll check it when the engine room's dry. I can't pump both at the same time.'

In the engine room I could see smoke and water being blown out from one of the cylinder heads. I cut off the diesel supply to that cylinder; the engine ran slower but it wouldn't do any more damage. When the water level was safely down, I changed over the pump to drain the hold. There was no water in it and I switched it back to the engine room to dry it out completely. Using hammer and nails, I made a more permanent repair on the engine-room hatch and then joined Simon in the wheelhouse.

I looked at the clock: an hour had passed since our crisis. Simon was soaked and shivering with cold. At least I had been in the shelter of the engine room. The boat was now managing easily as she ran before the wind, and I thought the sea had moderated: it was rough, but unless we were hit again by a rogue wave we should be OK.

'I'll alter course to Fraserburgh, Sy. It's the closest port to repair the engine.'

I drew a line on the chart and he turned the boat to the new course.

'I'll put the kettle on and get changed,' he said, taking off his oilskin jacket.

Looking forward at the sea ahead, I wondered what I was doing here, oily water dripping off my trousers and running across the wheelhouse floor, forming little streams as it first ran one way and then the other, following the roll of the boat. This wasn't the right way to take our scrap south. I saw in that moment that we were far too small and heavily loaded to be going this distance.

I was relieved to get ashore at Fraserburgh, just to stand on the pier and feel the reassurance of firm land beneath my feet. But there was little time to relax, as the engineering company the harbour had contacted after I spoke to them on the radio came immediately and wasted no time in starting the repair on the engine. Neither of us wanted to go to sea again in bad weather, but with no certification we could have been held up in Fraserburgh for carrying cargo. *The next boat we purchase must have some form of cargo certification,* I thought, as we left Fraserburgh. Fortunately we had a good run down the coast to Tayport, being partially sheltered by the land.

Unloading the boat in the peace of Tayport – our home port, which had no officials – we left the propeller blade and other scrap in Fife while we went with a lorry load to Glasgow. Travelling as passengers in the truck, we kept an eye on our hard-earned scrap and helped to sort through it in the yard of R.M. Easdale & Co. Ltd, a metal merchant and foundry. It was a slow process, as the scrap was divided into the various categories – copper, gunmetal, yellow brass, lead and white metal – all to be weighed and sold at different prices. Pipes and pieces with enclosed areas were looked at and banged on the ground to make sure they did not contain dirt, as the merchant did not like to pay copper prices for sand. His large warehouse had security like Fort Knox but we were told it still suffered from the odd break-in and to avoid losing anything from his nice old-fashioned, wood-lined office there was absolutely nothing of value in it – the metal prices being relayed to him from the international markets came from an old black-and-white television.

Metal prices fluctuated daily, and on occasions in the future if we thought the price was going to drop we would agree the price on Foula and Lindsay Easdale would stick to that price when he paid us a few weeks later. Even when the price dropped considerably he always stuck to the verbal agreement. The whole island began following the copper prices and asking us what we were going to do when they heard they had fallen.

Finally, with a cheque in hand, we were off to the bank to pay it in and draw out a few pounds to have in our grubby, scrap-soiled hands.

While in Fife, Peter, from my days on Barra, was ashore and wanted to meet us. I had kept in touch with him – letters were easily written during the evenings on Foula, with no television or other distractions. When we met him, he looked well, with his new-found wealth from commercial diving. I looked down at his feet, instinctively looking for Yak, his Barra sheepdog, but she wasn't there. Peter realised who I was looking for.

'Yak's got a good home,' he said.

'Good,' I answered and meant it. Our correspondence had cleared any clouds from our time on Barra. I was pleased he was successful in the world of oil-rig diving. While his fiancée started chatting to Simon, Peter honed in on me.

'Are you sure you don't want to go offshore?' he asked. 'It's easy money.'

'No,' I replied. The money was good, yes, but I was looking for more than just the money. Now I was on Foula, it was easy to refuse. I had no doubts about my future.

'I'll come to Foula to dive with you as soon as I get some time off,' he said. 'Some easy diving for a change!' Peter laughed.

He was in for a surprise.

Returning to Shetland with *Trygg*, all thoughts of bad trips were behind us. We were lying patiently alongside Blackness pier at Scalloway after refuelling and loading goods, getting ready to return to Foula as soon as the weather improved. Sitting in the wheelhouse, we were debating whether it was time to wander up to the pub when we were distracted by two girls shouting from the pier. I lowered the wheelhouse window.

'Can we help?' I asked.

'We've been camping in Shetland and we'd like to visit Foula for a few months,' said the taller one. 'We've been told you live there.'

'Yes,' I replied. 'How can we help?'

'Is there a regular ferry service?' she asked.

'There's the mailboat,' said Simon, who had lowered the other wheelhouse window as the conversation had become interesting. 'It's meant to leave once a week from Walls in the summer.' We were not a passenger-carrying boat and were careful not to take the mailboat trade, other than helping out, which seemed to be most of the time at that moment. It made no difference to us, as we never charged.

'Can you tell us about Foula?' came the reply.

'Fancy going to the pub?' asked Simon.

We walked up the road to the Scalloway Hotel. Lesley, the taller of the girls, walked in a positive manner, something I had noticed with many of the walkers on Foula. She shook her dark hair out of her waterproof jacket, where it fell straight and neatly to about six inches below her shoulders. Roz, her friend, was slightly shorter, her walking less coordinated, her clothes smarter. She had hooked her brown hair behind one ear, and had a slightly nervous smile. She appeared more emotional – someone who would never be a good poker player. They both looked happy, sunburnt and fit.

'Is camping our only option?' Lesley asked.

'Yes,' said Simon. 'Elizabeth, our neighbour, has the only available accommodation, a small two-roomed cottage, but it's usually booked by Germans for most of the summer.'

Simon and I looked at each other. At times I thought we were like an old married couple the way we were quick to pick up on similar thoughts. This seemed a perfect opportunity to find a replacement for Deidre, who had returned south to work. There was no need to say anything to Simon; I could sense him working towards it. I wondered which, if either of them, might be interested in the job.

After hearing about their travels through Shetland, their time at university and Lesley's life in Birmingham, I doubted that either of them would want the work. They appeared too independent, too organised, too city-orientated. Lesley had studied maths at Cambridge University – not that we thought that was a qualification for our cook. By the time the bar closed, they both wanted to stay on Foula next season. The arrangements for their return to Shetland and then the trip to Foula were easily agreed, as it had worked well with Deidre. I then wondered who got the best deal. I think they did.

Roz arranged to come ahead of Lesley at the beginning of the following summer. When the time arrived she travelled to Walls on the Shetland mainland, where she expected to catch the Foula mailboat. The boat was a day late, and then another delay in Walls, so it was 6.40 p.m. before they left. On Foula the visibility had only been about half a mile all day. Half an hour before the boat was due, people as usual started collecting at the pier and in our house. We placed a Tilley lamp in our window in the hope that the light might help guide the boat in. After a long wait

Harry went to find out if there had been any word from Walls, only to return with nothing to tell us. The boat was now more than two hours overdue.

He called out the Aith lifeboat on the west of Shetland, the most northerly lifeboat in the UK. The engine in the mailboat had been temperamental for a long time and the most likely explanation was that she had broken down. As she had no radio, there was no way to communicate with them. *Trygg* was on a mooring in East Voe at Scalloway, and we could not go out in the inflatable or we'd get lost. We took a torch and walked up the coast just in case they had come ashore on the island.

I thought of Roz. We hardly knew her, but we felt responsible that she was at sea. I had every confidence in Jim and knew that he would take no risks. He was a good mechanic and, although it might take him a while, he would fix the engine if it could be fixed.

It was three in the morning when we saw a torch coming towards the Haa.

'They've had to return to Scalloway,' Harry said, as he stopped to chat to us.

'What happened?' asked Simon.

'They thought they'd missed the island and turned back,' he replied.

Elizabeth was down early the next morning, hurrying to tell us that her friend had phoned to say that one of the newspaper headlines was 'Foula Mailboat Missing – With Divers' Cook on Board'.

Roz arrived two days later, none the worse for her adventure. In fact, she was more buoyed up than concerned.

Ten days later when the mailboat left on its run, it only got half a mile before it broke down. We watched it for ten minutes with the binoculars before setting out in the inflatable to see if we could help. It was decided the engine needed lengthy repairs.

We towed the boat back with the inflatable. It managed slowly but surprisingly well. This was the day Lesley was due to arrive. It was beautiful, flat calm, with perfect visibility, and I set off in the inflatable to pick her up at Walls. The trip took less than an hour.

At Walls, the mailbags had been made ready, so it seemed sensible to take them over: common sense still triumphed over regulations, although I did joke with the postmaster that we were so much faster than the mailboat that we should only be carrying first-class mail. Lesley and I stuffed the bags into plastic sacks to keep them dry and set off for a quick return trip.

Back on Foula, we were met at the pier by Harry, Elizabeth, Jim, John-Andrew, Roz and Simon. As expected, Lesley and Roz received a ribbing for bringing bad luck to the ship, though any calls of Jonah were made in jest!

9

Visitors to Foula

The two complete *Oceanic* propellers weighed twenty-nine tons each and were still attached to the ship by the twenty-four-inch-diameter propeller shafts. Each propeller turned in a different direction, one right-handed and one left-handed, requiring differently shaped blades. Consequently, the ship carried six spare blades at six tons each, although we only lifted five. One spare blade was unaccounted for until we received a letter from an old lady who told us that the *Oceanic* had lost a propeller blade when she was travelling across the Atlantic.

The central boss to which the three blades were bolted weighed twelve and a half tons of gun metal. As an indication of value, Simon had been looking for a flat in St Andrews and each propeller at that time was worth two flats. There was no way we would leave them, although they would be difficult to recover, as it would be a near-impossible task to unscrew the large nuts that held them on and then pull them off the shafts. Our only option was to cut the shafts, remove the blades and lift the bosses with a small length of shaft within them, making our maximum lift about 14 tons. We hoped – but were not sure – this was possible with *Trygg*. The starboard propeller was resting on the seabed, the two lower blades partly buried in rock and wreckage; the top blade had been broken off. After working through three tides, Simon, John-Andrew and I had physically dug out sufficient material from beneath the starboard shaft to place an explosive charge of 100 lbs of submarine blasting gelatine. The intention was to break the remaining two blades off, cut the shaft and remove the nut and spinner, a cone-shaped cap that fitted over the nut. The port propeller was more deeply buried and it took seven tides for us to tunnel beneath the shaft, using small charges of explosives to break rocks and free wreckage. There was the continual danger of debris falling on the diver and trapping him while we excavated, but now John-Andrew had joined us, two of us could be there, the second diver removing the debris as it was pushed back inside the tunnel by the first diver. Eventually the charge of 100 lbs was placed at the head of the excavation and the electric cable paid out and connected to the detonator. The inflatable was manoeuvred well clear of the area – with large charges there was always a strong shock wave felt on the bottom of the boat and we would expect the water above the propeller to be thrown into the air. Opening the lid of a small plastic Tupperware box to reveal a battery, we took the two bare wires at the end of the detonator cable and touched them on the spring terminals.

The explosion was instantaneous, the shock going through our feet and lifting the fuel tank and anchor off the wooden floor, making us feel as if the inflatable had been lifted into the air. Within a fraction of a second, the water above the propeller erupted, a plume like a nuclear explosion rising into the air and obscuring the island from view. This was followed by large volumes of gas, the beautiful clear water turned dirty with kelp and oarweed as it was pushed to the surface, turbulent water continuing to push up from below.

We reeled in the detonator cable and watched as a large number of fish came to the surface. Usually there was only a handful, if any, but this was the largest charge used on the wreck. The seabirds were determined to take their share but we collected as many as we could.

Although elated by the blast, we would have to wait until the next slack tide before we could see how effective it had been; the present visibility would be zero and the tide would soon be too strong for us to dive.

Lesley and Roz, instead of going for their usual walk, had watched for our return after hearing the bang. Our hearts lifted as we saw them running down the pier to take the ropes: Roz with arms and legs everywhere, Lesley more contained. Other islanders turned up too, to find out if we'd been successful. It was nice to be welcomed back, the adrenaline still buzzing round our bodies after the blast. Once the boat was secured, we passed up the unused explosives, the battery, diving bottles and empty fuel cans before we all walked up to the Haa. Sitting in front of the roaring peat fire, we thawed out and discussed the day's work, the explosion never far from our minds, my heart still thumping in my chest.

The charge was completely successful on the starboard propeller, but on the port side it failed to remove one of the blades and an additional small charge had to be used to finish the work. The propellers were now ready for lifting, each of them in four pieces.

But, hoping to lift the propellers immediately, we were disappointed when gales set in. Lesley and Roz started painting the house as Simon and I went off with Harry, Jim, Ken and John-Andrew to cut peats for Elizabeth; it was traditional that the peats were cut for those unable to do it for themselves. Elizabeth, in return, gave all the peat cutters a big lunch and tea.

During the next few days we attempted to complete an endless list of jobs for ourselves and Elizabeth, becoming as busy as when we worked the wreck. We looked forward to good weather to get back on site, our only break from island work occurred when we were called out to clear nets from the propellers of two Shetland fishing boats working off Foula. We never charged but were always given fish, which added to our stock of fresh food. John-Andrew continued to bring us wild birds and we often had a bucket full of eggs under the sink. John-Andrew, for all his bird-hunting, also had a soft side. One day, unnoticed by him, we watched him gently protecting an eider duck on her trip down from the moor to the sea with all her chicks as the Great Skuas attempted to pick up and kill her young.

Lesley was a natural cook and Roz learned fast. With no electricity and a limited supply of running water from our new tank on the side of the house, cooking could be a challenge to those not used to it. There was some initial disappointment from Roz when, much to Simon's enjoyment, he broke a knife on one of her scones. It had more to do with the Haa's old knife than the scone, but the event was rarely forgotten. They contributed far more than their cooking skills; it felt like we were all part of a family, our door was always open. Islanders, fishermen and visitors came in to be welcomed, often fed and entertained by Lesley and Roz, whether we were there or not. It was a shared house, everyone liked them: they were fun, cheerful and full of buoyant energy. Despite having no television, only a radio used almost exclusively for the shipping forecast, or sport in Simon's case, entertainment was rarely a problem. Simon and I had a pile of books for quiet moments. The girls had their own books from university that became a useful source of material for teasing them. We would read obscure passages from *Town and Country Planning* or *Child Psychology* and ask one of them to explain it. We soon discovered it was a dangerous ploy, as they were quick-witted and easily scored points off us, and at times the conversation was like wading out into deep water. Simon, if defeated, would relent with the words 'just probing for weak spots'. Here were two clever people who would go off to some city and get top jobs; well beyond anything Simon and I might aspire to if we were not in salvage.

The girls would often help Elizabeth, who would come down to the Haa afterwards and apologise, saying, 'Boys, boys, I had to borrow your cooks, I was needin' a hand wi' my sheep and there's no one else.' They were always pleased to help and when they heard by chance that it was old Tom's birthday they baked a birthday cake, placing a small candle in the middle as none of us knew his age. When they gave it to him, he put his pipe in his pocket before gently cupping the cake between his hands. Almost bereft of words he walked up to his lonely croft house shaking his head, a tear trickling down his unshaven face.

Peter had arranged to come out to join us with his new wife when he was next ashore. He was earning an enormous day rate but keen to dive on the *Oceanic*. I could see he was unable to let wreck diving slip from his life, and no wonder: it was like a drug, the sheer adrenaline rush when you saw those large structures underwater. It certainly was to me. Several months previously, when Simon and I had been sitting in a pub with him, he had explained his financial position.

'I'm making a lot of money now, but I'm worse off than when I was diving for sea urchins in the Firth of Forth. I had to dive for the urchins when I couldn't make money at salvage with Dan. I sold them from the back of my van to tourists.'

Simon and I laughed. 'How can that possibly be?'

'It's simple,' he said. 'I always spend more than I earn. When I was diving for sea urchins, I made practically nothing, but people bought me pints of beer and often fed me. I couldn't borrow from banks, so I borrowed a few pounds from friends to keep me going.'

'But you're making a fortune now,' said Simon

'Yes! But because I'm earning a lot of cash now, I'm the one buying the pints and meals. The bank wants to lend me money, which I foolishly take, and a vast debt appears to be rolling up in front of me. Financially I was better off diving for sea urchins.'

We knew Peter had spent extravagantly since his job offshore; he had also bought a house and loads of diving gear. He lent us his new inflatable boat when one of ours was damaged in a gale and we needed an immediate replacement. His inflatable gave us time, using it for a month before our new one arrived. He then refused all offers of payment. It just shows what goes around comes around. Simon was surprised he had not insisted on a payment, but I wasn't. Two dozen red roses came to mind.

Peter and his wife only intended to be on Foula for a few days and arrived by chartering the plane. They appeared very happy together, and Peter was keen to dive on the *Oceanic* while his wife wandered round the island. This allowed us to continue working that day, taking Peter with us as an extra diver. Simon explained the diving conditions. Although he had never dived with Peter, his reputation preceded him through my stories. I had no concerns about taking him out, although Peter was sceptical about the hazards of diving on the reef, tactfully implying that it was our lack of skill, rather than the tides, which were at fault. After our previous experience of nearly capsizing a small fishing boat we were not going to change our plans but continue to make all heavy lifts at the bow.

'If the lift's too heavy over the side, we can drop it down again,' said Peter.

'No,' I said. 'We're only attaching pieces that are aft of the boat on her mooring, clear of wreckage and we know we can lift.'

Peter and I went out on *Trygg*, Simon and John-Andrew motoring alongside us with the inflatable. We waited close to the wreck for the mooring buoy to come up from being pulled under by the strength of the tide. When the tide initially eased, we moored up and sat on deck waiting for it to slacken further before it would be possible to work. It was not that it was dangerous; when the tide was too strong, the diver was simply swept away before he got down to the wreck and had to be picked up in the inflatable – something that had happened to Simon and I on many occasions until we had learned to gauge the tide correctly.

'Have you dived in this sort of tide before?' Simon asked.

'Of course,' said Peter in his usual positive manner. 'It's a piece of cake.'

'Are you OK diving first?' added Simon, while we were sitting on the deck waiting for the tide to ease.

'Absolutely,' replied Peter before looking at the water. 'It looks OK to me.'

'I would leave it a while,' I stated. 'It's still a bit strong.'

But Peter was determined; we helped him on with his gear and then tied a rope to the mooring that trailed back alongside *Trygg*; it was our usual routine, being impossible to swim forward to the mooring rope in the strong current. Peter took the rope and carried out his neat back flip into the water. When the current caught him he held onto the rope, steadying himself, before pulling himself hand over hand to *Trygg*'s mooring rope, where he disappeared below the surface.

'He's doing well,' remarked Simon, as we walked to the bow of the boat to watch him.

'Yes, maybe we're not as good as we think,' I answered. The underwater visibility was excellent, allowing us to see Peter as he pulled himself down the heavy rope. But suddenly there was an eruption of bubbles and he came quickly to the surface, still holding onto the rope.

'The current's too strong,' he spluttered. 'It's taken my mask off!'

'Maybe have to wait a few minutes,' said Simon. He turned his head and looked at me. 'Well, we're not so bad at diving after all!'

'Let's get him out of the water. It'll be at least another ten minutes before it's workable,' I said, as we smiled at each other.

After Peter had finished diving he stood on the deck shaking his head. 'It's fantastic. I've never seen anything like it. The size of those engines, they're like underwater cliffs, and the rudder's the size of a complete fishing boat. You'll have your work cut out getting all the brass off her,' he said. We needed no encouragement, but it was the opinion we'd hoped for.

Back on the island we heard that Peter's wife had been the subject of numerous photographs as she wandered round the Haa and the pier, being asked to pose by a photographer who was over for a few days. When he had finished his photographs, he took his notebook out and asked her how long she'd lived on the island. To which she gave the reply: 'I arrived this morning.'

Our diving was not restricted to Foula. At times we wanted a day or two's break and would look at wrecks on the Shetland mainland and Fair Isle. Before Lesley and Roz left us we took the boat to Fair Isle. I had recently broken two fingers on the *Oceanic*, after they had become jammed under a lift when placing a sling under it. As it made it difficult for me to work on the Shaalds, where both hands were required, it was an ideal time to take a busman's holiday and enjoy a complete change of scenery and some easy diving. During the winter the *Norseman's Bride*, a steel trawler from Orkney, had been lost on Fair Isle. She had been sheltering on the north side from a southerly gale when she ran ashore on one of the many reefs in the area. Towered over by high cliffs and in gale-force winds, it had not been easy to rescue the six crew members in the heavy swell. Initially an oil-related vessel had sent in a rubber inflatable and they had rescued three of the crew, but it was then blown over by a heavy gust of wind and had to be recovered by its base ship. The Fair Isle mailboat *Good Shepherd* was launched and, with the detailed knowledge of the area known by the island crew, they manoeuvred the boat close enough to the wreck to launch a small Shetland yoal from the deck and row in, recovering the three remaining crew.

We knew the owner of the *Norseman's Bride* and his brother, who had the *Bountiful*, a good-looking wooden trawler; they were both successful fishermen and had a reputation for fishing in the worst of weather conditions. A crew member had been telling us that one of the brothers had bought a cine camera with the intention of filming in bad weather. During a really bad gale the two boats went out just to film – one shot his nets while the other filmed them working.

'Of all the things to happen,' he said. 'We had a massive haul of big fish, the brothers immediately stopped filming and made us shoot the nets again in this appalling weather.' I wasn't surprised the trawler ended up on the rocks.

Stewart and Jerry from the island came out with us to guide *Trygg* as close as possible to the site of the wreck. Simon dived first, discovering she had completely broken up on a boulder seabed, but there were some bits of equipment worth recovering. The visibility and seabed scenery was fantastic: beautiful clear water where we could look down at the wreckage with bits of trawl net floating up, mimicking some form of exotic weed. Towards the cliffs there was a narrow beach made up of large boulders that were strewn with smaller pieces of wreckage that had been ground up by the stones moving in the winter gales. It was easy to tell what moved on the seabed, as it was stripped of weed during the winter, with new growth showing in the spring. The netting caught on the wreckage was not a hazard because of the superb visibility.

I unscrewed the nut off the propeller about halfway but forgot to tell John-Andrew that it was a left-handed thread. He went down and unknowingly tightened it up again! After the propeller was free we lifted it out, along with some sheaves and other bits of equipment we could use. Pleased with our work we went round to the site of the *Canadia* on the west side and used a small explosive charge to remove the large boulder from the top of the remaining propeller blade – the one that had defeated Chris.

On our return to Foula without the girls, the weather remained fair, but the Haa seemed quiet now that they had left. When I had time to think, I realised I missed them; they were an unknowing influence. We had become fond of them and used to them shaping our lives ashore; this different way of life had crept up on Simon and I as bachelors. I was twenty-seven and Simon two years older. Many of our friends were married. I remember hearing my aunt Jean saying, 'Who on earth is going to marry Alec?' She was probably right. I was still unsure how Foula would feature in my life after we had finished on the *Oceanic*. My mind was so clear on the salvage and yet confused in my personal life, one voice telling me I was fine and just get on with life, while the other was telling me there was more to life.

Simon, I knew, was happy. Far more relaxed and unconcerned with his social life, he accepted everything as it happened, rarely planning further forward. We were both making money, perhaps we were looked on by the older generation as in *Pride and Prejudice*: 'it is universally acknowledged that a single man in possession of a good fortune must be in want of a wife', although that had been written 178 years earlier. I'm not sure that Jane Austen had taken salvage and Foula into consideration.

The propeller bosses would be our largest lift, at an estimated 14 tons each, including the shaft and nut, which remained attached to them. This was more than twice the weight of the single pro-peller blades we had lifted. Fortunately, attaching one of the bosses went like clockwork, with Simon and John-Andrew making a quick job of it after the experience gained from attaching eleven propeller blades and the anchor in the same way. As the boss came off the seabed in the slight swell, the bow of the boat went down and down until we were not sure whether *Trygg* would take it. It bounced several times before we had the full weight and started to drift.

I shouted back to Simon, 'Is our prop still in the water?'

He looked over the stern. 'Can't see properly but most of it looks to be out.'

'Put her slow ahead,' I shouted back, as I watched the boss swinging beneath the bow while the tide pushed us clear of the Shaalds. I moved back from the bow; it was almost completely submerged and at times I could hear the propeller thrash as the stern lifted clear of the water. Just three miles to go, but it was going to take a long time.

On reaching the entrance to the voe, water was pouring in between planks at the bow that had never been below water and the angle was so extreme that I could not get suction on the pump. It was a repeat of our first lift and I went in with the inflatable to fetch some of the people watching us and then organised a chain of buckets. I knew we would never get anywhere near the crane with it, as it hung too deep. We would drop it near the mooring, making it easy for us to take the nut off and pull the short length of broken shaft out. Then we could worry about getting it to Scalloway. When the winch paid out and the boss gently touched the seabed *Trygg* gave an audible sigh of relief as the tension came off her. The leaking stopped and we tied her to the mooring, reorganising the pulley blocks in preparation for lifting the second boss.

We had been busy working all hours, eating badly with the girls gone, and sleeping whenever we could. We did not notice Jim return with his twenty-foot yoal, the boat he was using instead of the mailboat, which had not yet been repaired. It was a long, exposed trip for a small, open boat and we were surprised to hear that he had brought a girl to help Elizabeth on her croft.

On the next non-diving day, Simon remarked, 'Let's go up to Elizabeth's and see her new slave' – a term used by Simon for all the young hands Elizabeth managed to employ, always girls.

'She must be brave,' I said, 'to come over in that small boat.'

We pushed past the 'caddie' lambs that barred our way to the door, while Elizabeth fussed over us with her usual kindness. Moya, the new girl, was sitting on the resting chair next to old Joann, learning how to spin wool. Her hair was tied up with a piece of leather with a wooden pin through it, keeping it clear of a polo-neck jersey. Her jeans were dirty from the croft work; she looked young compared with some of Elizabeth's previous helpers. Simon discussed the latest island news with Elizabeth as I glanced at Moya. At times our eyes met. I knew from past experience that caution was required when drawing her employees into the discussion; she might be sent off on a small job, as Elizabeth liked to be the centre of attention. My thoughts were far from Simon's gossip when I asked Moya where she was from.

'The isle of Lewis,' she replied, before Elizabeth noticed and sent her on an errand.

'How'd you get your new help? asked Simon.

'Jim's brother, who's a schoolteacher, had met her waitressing in the Queens Hotel for a couple of weeks while she found out how to get to Foula. She'd been on St Kilda and wanted to visit the island.'

'How long's she here for?' I asked.

'She'll bide here for a while. Mind, I've plenty work for her.'

'She's been lucky to get across,' I said.

'Aye, the boat's been broken down for weeks, but the weather was good. Jim decided he would cross in his boat,' replied Elizabeth. 'Harry says she's just taking some time to see the islands. He says she'd left school at sixteen with enough qualifications to study medicine and is taking time off.'

'It must be the first time that the boat's turned up at the right time, on the correct day, at the Foula pier in Walls,' Simon stated, as if wanting to start on a new subject.

'Jim did well with his peerie yoal,' Elizabeth replied. 'But I hope he'll get the right boat fixed.'

On the way out we passed Moya with a sack of hen feed. She looked as though she liked the work.

'Enjoying Foula?' I asked.

'It's certainly different.'

'You were lucky getting to the island.'

'When I saw the size of the boat, I wasn't too sure,' she replied with a smile.

Walking back to the Haa, I asked Simon, 'What do you think of her?'

'She's just another of Elizabeth's helpers, she'll last a few weeks,' he replied and looked at me with interest. I knew I had to be careful, any hint of weakness or affection and Simon would make fun of me. We knew each other too well.

While busy lifting large pieces off the *Oceanic* and dropping them in the voe I often saw Moya working in the distance. Bright and full of energy she appeared a useful hand, not like some of Elizabeth's previous helpers.

The Haa was a place most visitors migrated to at some stage of their stay on the island. When Moya came down I was out, but Simon and John-Andrew were there. On my return she was sitting in what was my chair – we had become creatures of habit! Simon, John-Andrew and I always sat in the same places, and most of the island knew this and found other chairs, but Moya didn't move, and Simon and John-Andrew started making comments just as a bit of fun. 'Alec's back hurts if he doesn't sit in that chair . . .' Followed by, 'Alec really likes to sit there . . .' But it had no effect. As they kept up the banter, she just smiled.

I sat on the floor with my back against the arm of the chair, pleased she had come in to see us. New people always added something. Her travels to St Kilda were interesting and it was while she was on that unpopulated island that she had heard about Foula, and became determined to see the most inaccessible inhabited island in Britain, as some people said it would be the next island to lose its population.

'Never,' I said. 'As long as there's a mailboat, the island will have a stable population.'

'What's a stable population?' she asked.

'I don't suppose there is one on a small island; they've all to find partners off the island, as they are all related.'

Unfortunately, just then Elizabeth pushed her head round the door to break up the conversation. 'Is ma lass here?' she asked. After that I could only snatch odd moments, as Moya repaired a fence or came down to the pier to fetch something from *Trygg* when we'd been out to the mainland. While *Trygg* was at the pier, I would lean out the wheelhouse window, organising the discharging of any cargo, and thought I could sense her watching me. She had an attractive smile, a smile with eye contact. Hard-working and bright, she had a good future before her; I knew she intended to go to university, so it was unlikely she would have any interest in me. I rarely worried about my appearance, but now I wished I had cleaner and smarter clothes.

'She's turned your head,' said Simon, watching me in the wheelhouse.

'Maybe, Sy. Maybe,' I replied, as I saw Simon give me a quick smirk.

After six weeks it came time for Moya to leave. Elizabeth was paying her so little that she had barely enough money to get home. I discussed it but she refused assistance, leaving my offer on the table as a last resort, but I wanted to help; it would have been an excuse to keep in touch. The issue was finally settled by old Tom, who paid her some money for chores she had done, saying, 'You'll be needin' this as yon ald wife'll not be payin' thee much.'

I considered myself a logical person, an engineer who considered all the evidence before making a decision. I thought I was determined but not impulsive, yet suddenly my life was changing. I was intrigued by this girl. There was no logic in this dream, other than the empathy I felt. I had no indication that she felt anything for me, but I missed her when she left. Unfortunately Simon realised it, using it at times to my disadvantage, although events started to take over our lives.

We were now in a rush to get the boat loaded before the weather broke. Visitors were still coming to the island, mostly blokes, but as the weather deteriorated, and there was a day of heavy rain and wind, we occasionally had someone knocking on our door, absolutely soaked to the bone and shivering with a tattered tent stuffed into the top of their backpack. They would huddle in front of our roaring peat fire before collapsing on a bed in our spare room as if they had had little or no sleep. The blokes were never as sociable as the girls. They usually came to the island for a particular reason, such as birds, geology or photography, being there *to do something*, like us. After warming up and a night's sleep, they took the first available opportunity to get off the island. I often wondered if they felt slightly humiliated at having to come to a house and ask for help. Generally the girls were more interested in the people and the way of life; they would always contribute by helping us or the islanders in some way.

Our luck was in when Karen and Jenni appeared on the island from nowhere. I had never seen two people so happy. It felt like a switch had been flicked. It was not that the Haa was ever unhappy, but the girls brought back the elusive fourth dimension that we had been missing, often a carefree tone that added perspective to our salvage and male-orientated life. Perhaps I was my own worst enemy: I was intoxicated with engineering, reading every book I could find, often talking about it and the salvage project when there were no girls about. It was my default position. In the Haa my main reference for dismantling the wreck was *Sotherns*

Verbal Notes and Sketches, A Manual of Marine Engineering. It was my bible.

Karen had a birthday celebration when she was with us, which livened us up. They made full use of the house, with its cupboards full of supplies, so much so that I was no longer sure who were the residents and who were the lodgers. After the party they started to cook for us on a regular basis; it was a great boost, as there were now four hard-working men to feed, with John-Andrew and Magnie, his cousin, helping on the boat. We soon discovered their cooking was not perfect: they failed to get the Yorkshire puddings to rise! The happiness they brought with them was infectious and spread through the team, giving us a big boost as we laboured hard to load *Trygg* for her final trip south of the season.

We knew we had to leave soon, as the longer-term weather forecast predicted bad weather approaching. First we had to go to Scalloway to refuel *Trygg*, and as the proper mailboat was still out of action while Jim rebuilt the engine we decided to take in stores and fuel for the island before going south, as we intended to leave *Trygg* in the safety of Tayport harbour for the winter. Loading eighty island sheep and numerous empty 45-gallon barrels belonging to the islanders, we left the pier at 9 a.m. with Karen and Jenni, plus two islanders, who came to help with the sheep. Although penned tightly on deck, two people had to stand among them at all times to pick up any that fell over due to the boat's motion. Otherwise they would be trampled to death. It was not a pleasant job.

There was always a happy atmosphere on the boat when we sailed to the Shetland mainland. It was not that anyone was unhappy on Foula: it was, I thought, the anticipation of a change, fresh sights, new people and in Simon's case the gathering and telling of news. I used the ship's radio to call up one of the boats we knew to ask about the weather and then Karen and Jenni wanted to speak to them. The time passed quickly, little thought being given to the poor weather report as we sat snug in the warmth of the wheelhouse.

A friendship had developed with many of the people in Scalloway who worked around the pier. There was always a hum of activity: fishing boats refuelling or discharging fish, lorries arriving with stores, nets being mended, the occasional tourist looking at the boats. Today, with sheep on deck, we were the attraction. After Simon had fended off questions by the onlookers, he went off shopping with the girls; I went around in my boiler suit with an oily rag in my pocket opening *Trygg*'s fuel tanks, filling drums with fuel, lifting supplies aboard and chatting to anyone who happened to be on the pier. Fifteen barrels were filled with paraffin – the main fuel on the island, for lamps and heaters – six barrels with diesel for tractors, along with several tons of fuel for the ship's tanks. I waited until the afternoon before the transport arrived for the sheep; I'm not sure who was happiest, me or the sheep when they finally left the boat. Sailing immediately for Walls, heading into a swell, the boat pitched heavily, with lumps of water coming over the bow and running all the way down the deck to pour out the scuppers at the stern. It saved us cleaning all the muck off the deck left by the sheep, but one of the girls became sick. Again I felt helpless, knowing from personal experience that there was nothing that could be done apart from checking they didn't fall over the side. I thought of our trip south, planned for the next day, and hoped the weather wouldn't get too bad.

At Walls the inflatable and outboard were unloaded and carried to the Foula hut for use when we returned from the south. Four barrels of petrol, fencing materials and anything left in the store for Foula were picked up, along with numerous boxes ordered by the islanders from A.K. Reid's, the general store. The islanders were playing safe, as they were not sure when the next boat would run.

As we arrived at the Foula pier, Harry and some islanders were waiting to help unload and then top up the hold with bits of brass that were on the pier. The following morning we loaded some of Elizabeth's sheep that were bound for Fair Isle, but while we did this a heavy swell started to break over the pier, forcing us to move *Trygg* to her mooring while Simon went in with the inflatable to take the last few sheep out. As he went round the pier a large wave broke over it, flooding the inflatable and filling the tractor trailer with water, causing a rush by the islanders to prevent the floating sheep from getting over the sides. It was after two o'clock before we left with John-Andrew and Magnie as crew, along with additional passengers: the Crofters' Commission representative, his wife and two children, one island wife, and Karen and Jenni. They all planned to stay on Fair Isle. We had forewarned the Fair Islanders about taking over the representative from the Crofters' Commission in case he was coming across to inspect something; it was a courtesy we followed with all government officials that we carried. When we first purchased *Trygg* we had taken passengers with a slight reluctance, but our attitude changed when the mailboat broke down, as transport from Foula was so limited that people were desperate to get lifts off or on to the island.

'I'll be glad when we get rid of the sheep,' said Simon.

'How'd you think Elizabeth's sheep will like Fair Isle?' I asked.

'They're going on Sheep Rock,' he replied.

'They'll be lucky if the sheep don't jump off the cliffs when they try and catch them,' I joked. I knew, from experience, that the sheep from the Foula hill were wild. Many, like the ones we were carrying, were moorit coloured – a reddish brown. The sheep had achieved the status that the New Zealanders sought: they successfully lambed on the hill without any human intervention. If the weather was good enough, the Fair Islanders would take the sheep by boat to Sheep Rock, which contained eleven acres of grazing, almost an island, with all its sides being sea cliffs except for a collapsed ridge that connected it to the Fair Isle mainland. A chain had been secured in the rock so that it could be climbed from the seaward side. The sheep were hauled up from a boat using a rope. This piece of land, as long as it had sheep on it, gave the landowner, a shopkeeper, crofting status, allowing him to apply for government grants.

We all took it in turn to stand among the sheep, a job that required a bit of strength as well as a good pair of oilskin leggings. Miles from land, most of the passengers enjoyed taking a turn at the wheel, with the galley continually busy producing egg butties and cups of coffee. I don't think we could have given our passengers a more interesting trip if they had been on a cruise ship.

Arriving at Fair Isle we were met by Stewart, who quickly arranged to get the sheep off. It was a pitch-black night before we left, but with the help of the fishing boat *Dauntless* that lay at the pier beside us we manoeuvred out of the voe while he used his searchlight to keep us clear of the rocks. The number aboard *Trygg* had been reduced to Simon, John-Andrew, Magnie and me. It was a perfect night for steaming south, with a light following sea helping us on our way. As soon as we arrived at Tayport, I posted a letter to Moya.

When the winter came and we left Foula, planning to return the following year, I took time to fly to California. A Captain Fasold owned a salvage company there and had asked me to dive and give my opinion on a shipwreck, the SS *City of Rio de Janeiro*, which contained silver. Simon was busy completing the purchase of a flat in St Andrews and did not fancy coming with me.

Arriving at San Francisco, I recognised Dave Fasold from the salvage photographs he'd sent me. Throwing my kitbag in the back of his truck, we set off to his boat. Conversation in the truck was easy. He gave a commentary on the buildings and history of San Francisco, discussed the salvage projects he had been involved in and the ones he dreamed of being involved in. I sat there quietly, trying to take it all in. His salvage boat was a tank-landing craft bought from the US Navy; she lay alongside a wooden pier that was being demolished – his day job involved removing the piles, the giant wooden pillars, supporting the deck of the pier. They were made from long, straight lengths of pine, three feet in diameter, 50 feet long and were in good condition, except around the waterline, where a hundred years of use had shown some wear. It seemed a travesty to throw these magnificent tree trunks away.

The wreck was a venture that he and his friend Kirk had dreamed of. They had built a magnetometer, which would be towed behind the salvage boat and hopefully pick up the magnetic field of the wreck in order to locate it. I looked at the magnetometer as it lay in the aft end of the wheelhouse. It was new to me but I could understand the principle. This was to be its trial run. On the chart table was a picture of the wreck in better times. The *City of Rio de Janeiro*, an iron-hulled, steam-powered passenger ship, had been launched in 1878. She was lost in February 1901 when she struck rocks in heavy fog at Fort Point, close to the south end of the Golden Gate Bridge. The bridge was constructed at a later date and opened in 1937. The ship sank within a quarter of an hour, with the loss of seventy-two passengers and thirty-two crew out of the 201 persons aboard. She was sailing from Hong Kong and the insurance companies paid out £100,000 on the loss of the coin and silver she was believed to be carrying, which equates to £6,000,000 today. We discussed the possible location of the silver while we enthused about the prospect of our success. She had struck the rocks, so I was expecting an easy dive in beauti-fully clear shallow water, and looked at Dave as he talked, trying to work out his motives. In projects like this, the rewards could be enormous if successful, but this added an element of greed. The money was sometimes the sole attraction, but to me it was important that people were motivated by the sheer challenge of the venture, for it to be carried out at a reasonable cost. Not that there was any cost at this stage, it was just a bit of fun.

At the weekend, we left the pier and sailed out of San Francisco Bay, leaving Alcatraz Island to our north. When we passed under the Golden Gate Bridge the current swirled around us. I was used to strong currents, but this water was deep and I could see the swirls carrying sediment, creating poor visibility. I was nervous of entering the water. Realising that neither Dave nor Kirk had any intention of diving on this trip I became curious as to what tales he had been told about me and my capabilities. I was beginning to think they would both be disappointed. I had no superhuman powers. Peter or Simon would have been better at searching for a wreck than me; it was the dismantling and recovery side that I really enjoyed and that was my strength.

But then I relaxed as I realised most of the morning would be spent setting up the magnetometer. The presence of the Golden Gate Bridge was affecting it, which resulted in too many 'hits'. We were losing the day, so we returned to the most likely mark.

Now was the time for diving. I felt anxious using borrowed equipment, as I didn't know how reliable it would be, and it felt different, not giving me the reassurance I felt with my own gear. When the engines stopped I looked over the side with apprehension. I had come all this way: pride and commitment applied their pressure. I had to dive. *But,* I thought, *this is how accidents happen.* However, there was no time for hesitation.

With a coiled-up line attached to a small marker buoy, I made a quick flip over the side and swam down into the dark, swirling, unwelcoming water. My breathing was fast because I was nervous. I sucked hard on the demand valve in an attempt to extract enough air. Initially the underwater visibility was about 15 feet as I headed towards the seabed. Passing the 40-foot mark, the visibility decreased. It was like a thunderstorm coming towards me; the water became turbulent and I could feel my body being tossed around. I no longer knew which way was up – even my bubbles were dispersed in all directions. Realising I had no control I looked at my depth gauge, but the visibility was so bad that I could not read it. The depth felt stable, or perhaps that was wishful thinking. I went through my danger ritual, telling myself to relax: *Just relax, Alec. Just relax*. Repeating the mantra, I tried to work out the direction of the surface and safety. As quickly as the turbulence had started, it ended; I drifted out of it and the visibility cleared. I made my way to the surface and the boat came towards me. They pulled me aboard, the marker line twisted around my body, then I sat on the deck recovering, feeling that I had let them down.

My mind went back to Foula and a memory I had of Moya hanging on to the side of a boat in the voe. She had been swimming and was obviously breathless, fighting for air, not having realised how cold the water would be but lucky to find the small boat to hang on to. I was not cold, but I was still breathless, the adrenaline pumping through my body. I thought to myself, 'What the hell am I doing here?' and shook my head. All further dives were carried out swimming down an anchored rope. We never found the wreck. It had probably slipped into deep water.

Near the end of my stay, I found a letter waiting for me on the table in the lounge. I turned it over in my hands: it had been sent from Stornoway on the isle of Lewis. All thoughts of salvage were gone. I opened the envelope. Ever since I had written a letter to Moya I had been hoping for her reply. At home I had rushed down to pick up the mail, only to be disappointed. Now I would know. I tore the letter out, the two pages of airmail paper gripped carefully between my fingers. Laying the envelope on the table I started to read.

I thought of Moya sitting on my chair in the Haa, that curious smile. Yes, this was her letter. I could imagine her writing it and knew her writing as I had watched her making shopping lists for Elizabeth. After reading it several times I put it back in the envelope, got up and walked around the empty room, my hands moving as if I was trying to explain something. I was restless, impotent to act on the letter but needing to do something. I read it again. An immediate reply was not required. It was light and amusing, wishing she was out here with me. Each time I read it I imagined the meaning to be slightly different, but from that moment I knew I must see her again.

10

Life Changes

Arriving home a week before Christmas, I sent a short letter to Moya saying I would arrive on Lewis on 7 January. I had made up a reason, saying I intended to research some wrecks at Stornaway. These wrecks had been lost on Rockall, a distant isolated rock and reef that lay out in the Atlantic, west of Lewis. I lingered in hope of a reply as the days passed.

The ferry to Stornoway, the largest town on Lewis, left from Ullapool on the north-west coast of Scotland. It was a drive of 200 miles from Fife. I took my motorbike, as I knew I could guarantee to get it on the ferry, regardless of the number of vehicles aboard. There was no sign of snow, but gale-force winds and heavy rain soon penetrated my clothes, the rain trickling down the back of my neck. The road to Ullapool wound among the hills and lochs, causing sudden gusts of wind that tried to unseat me as I kept my speed up. I looked forward to the comfort of the ferry and some warm food. At Ullapool the ferry had been delayed because of storm-force winds but they let me drive the motorbike aboard. I settled on the boat, thawing out my painfully frozen fingers while I emptied my saturated backpack, hoping all my fresh clothes were still dry. Spreading my leather jacket and trousers over a radiator to dry, I fell asleep on a chair and thought how lucky Moya had been in getting to Foula. If Jim had not taken his own boat, she might never have got to the island. In my heart I knew this wasn't true: she would have got there somehow, she was that type of person, and that's what I liked about her.

It was the evening before the ferry sailed. The weather remained poor and I had to hold tightly to my seat in the heavy swell, wishing the three-and-a-half-hour trip would quickly pass. I tried to imagine her house, and what I would say. I knew nothing about her parents and it had been five months and two weeks since I'd last seen her.

When the motion reduced, I went on deck to watch our approach to Stornoway. It was 10.30 p.m. and pitch black with heavy rain, but the wind had lost much of its force.

I had 35 miles to go to the west side of the island on unknown roads. I tucked my map inside my jacket when the seaman waved me off and I roared up the ramp, the noise reverberating off the sides of the ferry and the heat of the engine already warming my legs. Just as I got to the top of the ramp, however, I saw Moya waving; I spent the rest of the trip in the warmth of a van with the motorbike safely secured in the back.

Valtos is a small village. In the darkness I was able to make out one row of houses on either side of the road before we stopped. Ushered in, I was shown my room on the ground floor – Moya's room that she had vacated. The bed was heaped with blankets and a hot water bottle beneath them. As I went to bed I could hear a slight whistling sound and looked at the window. A wooden frame had been tacked over the window, holding a sheet of transparent plastic to reduce the drafts from the old frames. In it were two pin holes, the source of the whistle. I smiled. This was like Foula.

The following day was the sort only seen in Scotland. The wind had dropped to a breeze, the low sun shone through the lessening cloud, and Valtos looked beautiful. At the bottom of the village was a small pier. We wandered down before walking the short distance to the beach – a mile-long arc of shell sand, well sheltered from the prevailing wind. It lay before us completely deserted, inviting us to enjoy it. We moved from the softer sand at the top to the firmer footing near the water's edge, where the small waves broke at our feet. The gentle sound made talking easy. Pauses went unnoticed and I hoped the beach would last forever.

For lunch we took my motorbike the three miles to Miavaig, where her parents used the old school buildings as the shore base for an oyster farm. We looked at the oysters, strung from rafts in a sea loch nearby. Interested to see more, I was shown the shed alongside the water's edge where the oyster spat were grown prior to being transferred to rafts. Moya temporarily worked in this shed, cleaning and tending the young oysters, in the cold, water-saturated atmosphere. On my final day we went to the cliffs on the extreme west and sat on a rock above the thundering waves. The sea was magnificent, exuding power and unpredictability as it changed shape and form, turning from hard blue water to foam and mist. This was life, real life, impressive, nature at its best, unfettered by human hand. Far out, more than 40 miles away to the west, lay the uninhabited island of St Kilda, a similar distance as Foula to Fair Isle. Two hundred and forty nautical miles to our west lay the lonely rock stack named Rockall.

Above the noise of the breaking waves we talked about the *Oceanic*; Moya mentioned university next year and then a silence fell over us. I knew I would soon have to make a decision. I said nothing but felt an echo from Foula, watching the birds gliding in the updraft from the cliffs. I didn't want to spoil the moment.

Ten minutes early, but I constantly glanced at my watch as I looked out for the incoming train. When it arrived, I swept my eyes along the doors. Moya saw me and stumbled, nearly falling onto the platform. Smartly dressed, with brown leather boots and matching jersey, she looked fantastic. I could not believe how lucky I was as we made our way out of Dundee station and drove down to *Trygg* at Tayport harbour, where she was being prepared for the next season.

At the pub for lunch, Simon, never able to resist a good story, told us that he had had a visit from the police at the flat he bought in St Andrews. The flat was on the ground floor, his bedroom looked onto the pavement. He had never been the tidiest of people and recently an old lady had looked in the window through the partly opened curtain and seen the mess in his bedroom. She had phoned the police to report that the house had been 'done over'. The police turned up and burst into the unlocked flat to find Simon sitting in his pyjamas and bobble hat, watching the television at eleven o'clock in the morning. The story relaxed me. Simon was happy, but I wondered how he would feel about Moya joining us on Foula – if she agreed to come.

Returning to *Trygg* in the early evening and parking among the bundles of timber that had been discharged from a cargo ship, we looked down at her, lying 14 feet below us in Tayport harbour. It was low water. We went down the ladder to board her, turning the heater off and shutting the boat up for the night. I watched Moya leap several feet from the boat to the rusty steel rungs set in the stone wall before neatly climbing up, as if she did it every day. It was a fine clear evening as we made our way round the stacks of timber to sit in the pick-up. I looked out of the window at the timber surrounding us. The pick-up engine was running to keep us warm. I switched it off and leant towards her.

'Moya,' I asked, 'will you marry me?'

'Of course I will,' she replied.

Moya left for Foula early, as she hoped to cross over to the island by fishing boat and get the house dry for our arrival. *Trygg*'s hold was filled with food for us and included orders from the island that we had bought at the cash and carry. We loaded our Morris 1000 pick-up and a three-wheeler car John-Andrew had bought, placing them either side of the hatch. Finally we waited for half a ton of explosives to be delivered before setting off for the island. Harry had warned us that this was early, but we were kicking our heels and itching to get back to the wreck. We had the usual blocked fuel filters at the start of the trip, as *Trygg*'s motion stirred up the fuel tanks, and there were minor problems with the radar. Otherwise it was a reasonable trip until we hit a northerly gale just south of Shetland. Unable to land at the pier on Foula, we went to Scalloway, where word got round that we were bound for Foula, resulting in a deluge of boxes and parcels to take to the island.

'The mailboat can't have been for a while,' Simon commented.

'If it's been like last night, I'm not surprised,' I said, realising Harry had been right and we should have waited for better weather.

Within a few days there was a lull, allowing us to make a run to Foula despite a gale forecast, but the weather conditions were not good at the pier, the boat ranging backwards and forwards on its ropes each time the heavy swell swept in. We lifted and drove the cars off before unloading all the food for the islanders, Jim making several trips with his tractor and trailer. Finally the explosives were unloaded and carried up to the Haa, Moya slightly shocked when she saw them placed in the cupboard under the stairs and the detonators in our bedroom.

It had been a long, exhausting day but we would have to take the boat to the safety of Scalloway if the weather continued to deteriorate. After a quick meal at the Haa, which Moya had prepared in advance, I went to bed to catch up on some sleep, leaving Simon to check the ropes and fenders on *Trygg* during the night. The next morning the weather was no better and I sailed for Scalloway with Moya, leaving Simon on Foula.

Moving out of the shelter of the land a long swell began to increase the rolling of the boat. Moya took over the wheel, while I went below to check the engine room. The trip would take three and a half hours and, with an imminent gale warning, I could not afford the engine to stop through negligence, with just the two of us aboard. I returned to take the wheel, both of us sitting in the wheelhouse enjoying the peace of being at sea after all the rush.

At times, life was not easy for Moya, as she had to contend with Simon and the business. The three of us got on well but Simon seldom passed up an opportunity for a little well-intentioned teasing. Moya proved very tolerant with us both.

'Don't peek, Moya!' Simon would shout as we ran naked past the kitchen door on our way to pick up warm towels after shedding our diving suits outside. Dressed and warm we sat down to eat and found her to be the best cook, Simon and I putting on so much weight that we had to limit her to one pudding a day. Even then Simon had to put gussets in his wetsuit, and, as the only spare tape we had was yellow, he ended up with a distinctive three-inch-wide yellow strip down the front of his suit.

Much of the summer involved loading the boat with scrap from the wreck and transporting it south. We were amazed at how much we were finding: as soon as we lifted one load, it exposed more. Moya learned quickly, working the winch on *Trygg* when John-Andrew had to tend to his croft. She also wanted to learn to dive but we were busy and I was unwilling to teach her, perhaps selfishly, but I knew how dangerous it was – although working on the boat proved to be more dangerous.

Going out to the wreck, John-Andrew and Simon took the inflatable and Moya and I went on *Trygg*. Simon often dived first, hooking on the brass to be lifted. I took the signals from Simon using a rope. The most dangerous moment approached as the brass came over the side; it could develop a swing as the boat rolled in the motion, so quick reactions were required by the winch operator to drop it in the hold or against the side of the hatch. John-Andrew had a 'feel' for anything mechanical and we both trusted him implicitly as a winch operator and diver. When Moya was not required on deck she would climb on the top of the wheelhouse to take photographs of the brass being winched aboard, often having to hang at crazy angles when we took a heavy lift over the side. The days flew by, Simon occasionally reminding me I was getting married at the end of the summer . . .

The date of the wedding had circulated through the island, but it was traditional that Simon as best man 'bid' everyone. He would go to each of the households and formally ask the occupants if they would be kind enough to come to the wedding. It took him a long time, as he would enjoy a cup of tea and a chat in most of the sixteen households. There were also islanders who worked on the Mainland, only returning for the summer break, increasing the total number of islanders to thirty-nine. Relations and friends were given a verbal invitation along with Shetland fishermen who worked around Foula, as they had become regular visitors to the Haa. The crew of the lobster boat *Juna* intended to come with their families.

'It'll have to be fine weather to land at the pier,' said Ian, the lobster fisherman.

'We don't want it too good,' joked Simon, 'or Alec'll be wanting to work on the wreck.' Simon, in his usual fashion, had spread the gossip that if it was a diving day we would have to fit the wedding in around our work.

With no regular organist, we asked Bobby Isbister to play. He was not a churchgoer but was a talented musician. The old harmonium in the church had limited notes due to the damp and various attacks by mice, so Bobby and Eric helped us take the harmonium from their house to the church. The mice had also eaten a few of its air pipes, resulting in the tunes played being dictated by what Bobby could play and the notes the mice had left intact.

With seventeen days to go, the minister from Walls, the parish that included Foula, phoned Elizabeth to try and contact us. I called him back as soon as Elizabeth gave me the message. It was a simple request for us to go over to see him, as Foula only had a missionary who was primarily the school teacher and had no additional qualifications, just the approval of the Church. We agreed to a meeting the following day, if the weather was suitable for a crossing. The quickest way was the inflatable, but the shipping forecast at six in the morning predicted gale warnings for our area. At the time the wind was just a light breeze from the east-north-east, with a heavy glassy swell. We had become accustomed to the weather patterns and decided the gale was unlikely to come before the evening.

Leaving the pier at 7.25 a.m., Moya and I were well protected in oilskins as we started on the trip to Walls. It wasn't wet, but the swell made a bumpy crossing, as the boat jumped off the top of one wave onto the next. Feeling exhilarated and confident after our fifty-five-minute run, we opened the plastic bags in which we stored dry clothes and tidied ourselves before walking up to the manse to meet the Church of Scotland minister. He was originally from Iceland, therefore used to islands and those who worked on the sea. Our salvage work proved a common interest, easing us into the subject of the wedding service. Ours was the first wedding in the Foula kirk for twenty-nine years, as the islanders tended to marry on the Mainland, returning to Foula in some cases to be blessed. He explained that, as a minister from the Shetland mainland was required, the weather had to be good enough for the aeroplane to bring him in. An island wag said later: 'If the minister is worth his salt, he could walk across.'

The warmth of the manse gave us no reason to hurry. I sat there wondering what he thought of us. We were happy, not taking it as seriously as we ought; it was an adventure to us, even the tradition of marriage. We left the manse with his blessing and no hint of any complications. Walking up to A.K. Reid's, the grocer's, we picked up our 'Waas box' and our mail from the post office before returning to the pier. Ahead of us Foula sat proudly on the horizon, with its rugged rocky outline looking strong and imposing. It was a home I longed to get back to, and the boat, as if knowing my thoughts, flew gracefully from one wave top to another; with the tide in our favour we arrived at the entrance to the voe five minutes faster than on the outward trip. Tying up at the pier at 10.30 a.m., the weather conditions were beginning to deteriorate. Simon, Jim and Rob met us at the pier and helped lift out the inflatable, but the sea conditions were now getting too bad to allow the mailboat a safe trip there and back. The gale was going to hit us.

As the time before the wedding reduced to days, Simon, John-Andrew and I continued diving, blasting tail-shaft liners, bearings and thrust blocks. We were taking every opportunity to lift small bits, with most of our attention focused on blasting the larger pieces in preparation for the next season. I was keen to use up all our explosives before the season ended otherwise they would be wasted, as we would have to dispose of them by burning.

With only eight days to go, the house had taken on the pleasant aroma of baking. Moya had finished her silk wedding dress, made in the Haa on an old treadle sewing machine, and was moving on to the catering. Our last day of good weather was five days before the wedding date. Two days before the wedding the wind was strong from the south-east. Fortunately it dropped, as it was a bad direction for any boat trying to get into the voe. The following day the plane managed in with Moya's parents, my mother and some of my relations, while two of my cousins working in

Shetland were dropped off by a fishing boat. Moya had tidied up the Haa as best she could, but my aunt came in and said, 'I see you haven't had time to clean the house. I'll make a start on it!' Moya was left open-mouthed.

On the day of the wedding the weather was too bad for diving, but with a light north-easterly wind there was no problem for boats at the pier or the aeroplane landing on the gravel strip. I had bought a new suit for the occasion, as Simon and I had only one suit between us, known as 'the company suit'. Everyone on the island had dressed for the occasion and we all looked like strangers. With no idea of the numbers coming, the small islander plane had to make another unplanned run that caused a slight delay – not that time was ever a problem on Foula. On the final run he brought the minister, the Reverend James Blair, who coincidentally had the same name as the navigating officer of the *Oceanic*.

I sat with Simon in the front. No one was concerned, least of all me, as Moya was already waiting with her father in the small vestry at the back. I looked around the tiny kirk. It was completely full. Built on a ridge that led up the side of Hamnafeld, the main hill in the centre of the island, it looked down over the gravel airstrip, which was just several minutes' walk for anyone arriving. We could hear the aircraft with its final load of passengers. I stood on my toes to look out of one of the plain glass windows but could only see the sky. The plane revved up as it turned on the runway, moving to the dropping-off point before there was silence as the engines stopped. A few minutes later the minister, the pilot and a few more guests came in. Simon gave me a nudge and a smile as Bobby, who had been improvising on the organ to fill in the time, was extending his repertoire to 'You Are My Sunshine'.

The minister came forward and then Moya was led down the aisle by her father, Bobby giving the organ as much volume as it would allow.

'Could you please sit?' asked the minister when the music stopped, leaving that total silence you find in the country, occasionally broken by a cry from a bird. I moved to sit down, but then Simon whispered, 'Not you!'

After the ceremony, we headed to the school. All the tables were moved back against the walls, as Harry and Jim warmed up their fiddles ready for the dancing. The Shaalds of Foula, a dance that mimicked the breaking waves where the *Oceanic* lay, was appropriately the first dance. With dancing in full swing those who were leaving by plane had no need to be concerned about waiting at the airstrip as the pilot was with us enjoying himself. At midnight we left the school, changed into our old clothes at the Haa before walking up to Loch cottage. Elizabeth suggested we use it for the night. Every spare bed in Foula was occupied, including our own, as the Haa was filled with relations. The walk was pleasant and peaceful after the dancing, and occasionally we heard wild hoots and laughter carrying across the island. It was a dry night with a light breeze. We looked down at the lobster boat *Juna* lying peacefully at the pier; three miles out from her lay the Shaalds and the wreck that had been instrumental in our meeting. At that moment the world held everything for us: we were happy, we were fit, we were young. The future could not have looked better.

11

Unexpected Storms

The following spring we returned to Foula, knowing the work on the *Oceanic* must come to an end, but we were all enjoying life on the island and the diving remained fantastic. The wreck still yielded brass but we could see it gradually being cleared, and now we had become used to the bad conditions we were able to squeeze in every available moment that could be worked.

There had been a strong wind from the west when we arrived. It was forecast to decrease but instead it had strengthened, while backing to the south-west. The original forecast made us think that *Trygg* would be sheltered on her moorings in the voe, but when the wind quickly increased in strength while backing further to the south, and finally to the south-east, we knew we had a problem. The wind was now driving the heavy Atlantic swell into the voe. *Trygg* was burying her bow deep into the waves before rearing up to strain at the chain that tethered her to the *Oceanic's* anchor. I knew she would not drag the anchor but the chain might break or pull itself out of its holding point on the boat; we could see it grinding where it was bent over the roller at the bow. The weather deteriorated further until we might have lost our lives making an attempt to board the boat and sail her out of the voe.

Moya, Simon, Harry and I stood in the shelter of the Haa as the darkness started to come down. I knew I had missed the opportunity six hours before to take the boat to Shetland, and I was regretting it. Now there was nothing we could do.

'Well, well,' Harry said, 'it's a foul night but at least you're all ashore.'

'Are you coming into the Haa?' I asked, as the rain increased in a heavy gust of wind.

'Na, na, I'll away and see if Elizabeth's needin' anything,' he replied, as he turned towards the road.

We went inside, pressing ourselves to the window, just able to see *Trygg* in the failing light when the broken water flashed white as it pounded against her. We eventually went to bed but during the night Moya woke me.

'Alec, quickly,' she cried, as she shook me before rushing back to the window.

'What's up?' I asked, as I hurried out of bed and crossed to be beside her. I could see little through the darkness, but there was a definite 'clink' to be heard, as if two metal plates were being banged together. We shouted to Simon while rushing down the stairs and out of the house to be met by an intense smell of diesel oil. Slowly moving from the shelter of the house, we watched the remains of *Trygg* being smashed to pieces on the rocks. There was nothing we could do except stand there, transfixed by the power of the sea as the enormous waves crashed against the shore covering us in spray. Simon joined us in his bobble hat and dressing gown, shaking his head as he looked at the place *Trygg* had been. Cold and soaked, we moved back towards the house, the violent wind making it difficult to keep our balance and the spray forcing us to hunch our backs to shed the water.

'Thank God no one was aboard,' said Simon.

'Wouldn't have had a chance,' I replied, as I looked back at the foaming water rolling fragments of the boat on the rocks.

'There's nothing we can do until daylight. I don't know about you, Alec, but right now I think I'd rather have a job as a bus conductor.' I knew what he meant. Simon had always been risk-averse.

We trudged back to the door of the Haa. 'I'd better tell Harry,' shouted Moya above the noise of the sea. 'He's up all hours and may see *Trygg*'s gone. I'll let him know everyone's OK.'

I thought of going instead of her, but I was so dispirited I just wanted to sit down and think. I caught her eye and nodded, watching her disappear into the darkness. The fire was stirred up and the kettle boiled as we sat in stunned silence, each with our own thoughts.

'I think I'll go to bed,' said Simon, 'if there's nothing I can do.'

'I'll wait for Moya,' I replied.

Moya eventually returned. It was two in the morning, but Harry had still been up. We went to bed but I remained awake for most of the night, unable to sleep as my mind mulled over our options. I finally came to a conclusion.

At first light I went outside, attempting to keep in the shelter of the Haa as I looked for the remains of *Trygg*. There was no sign of any large section of the boat. Moya and Simon joined me, the three of us venturing towards the pier to find it completely awash with broken bits of wood. The breaking sea prevented us from getting close to our rubber inflatable, which lay among the floating debris. Rigging ropes to hold onto, we attempted to recover it, the waves washing over us at times while Moya held onto the end of the rope that was threaded through a steel eye. Eventually I had to jump into the water to catch it, quickly followed by Simon, as Moya pulled us in. Once the inflatable was ashore we lifted it clear of the water and secured it against the gusting wind. Although I was soaked and cold I had found the task exhilarating – it had taken my mind of our immediate problems. We searched for *Trygg*'s fuel tanks, but there was no sign of them, or any indication of the diesel oil. It had left no marks and the smell had disappeared. I assumed it had evaporated. She only had a small quantity aboard, as we had not refuelled her since coming north.

Islanders came to commiserate, the house filled up, all pleased to be in the warmth of the Haa, sheltered from the wind and rain. Moya made drop scones on the girdle over the open fire, with catering tins of jam and thick-cut marmalade placed on the table for everyone to eat. The loss of *Trygg* was talked about in hushed voices, almost as if a person had been lost. The future was discussed while we listened to the forecast on the radio, gusts of wind up to 58 knots (67 mph) had been recorded at Sumburgh. We had been warned by the islanders that leaving a boat moored in the voe was a high risk and they had been proved right, but there was absolutely no sense of 'I told you so'; in fact, it was the very reverse, as they knew we would have had a problem working from Foula without a mooring in the voe. Harry as usual was concerned about our work continuing, and unusually but briefly came into the Haa. There were moments when I believed the loss was felt more by some of the islanders than by me. I knew I could do something about it, by buying a new boat, but for the first time I realised how fond we had all become of *Trygg*. Although an inanimate object, we had trusted her with our lives. Simon, of all of us, was the most upset by the loss. I had been fond of her but also looked on her as a tool, knowing that she could be replaced by something better now we had gained more experience and could pay more money. I had thought about it many times, but I knew Simon would prefer to keep *Trygg*.

Later that day, after the rain stopped and the wind eased, we walked up the side of the voe to look at the wood-strewn beach. 'Let's just lift what we can from the *Oceanic* with the inflatable and finish,' said Simon. 'We've failed. We can't work here with a big boat.'

216

'Of course we can,' I replied. 'We just have to be more careful with the weather.'

'No, I don't think we should get another boat,' he replied, visibly upset. 'Once we've finished tidying up, I'll get another job.'

There was no point arguing – at least he was happy to continue tidying up with the inflatable. We were near the peak of our recoveries; there was still more to come, it had seemed never-ending, exceeding our most optimistic estimates, but the work would become harder and require more skill. On the positive side, *Trygg* was insured and Moya, like me, was happy to continue, so we could afford to get another boat. I quietly thumbed through the advertisements in old copies of the *Fishing News*. It was a paper that had puzzled Moya, as in Shetland it was pronounced the 'Fashion News' and at one time she had wondered why these fishermen were going into the newsagent's and asking for it.

Simon was joined by John-Andrew in thinking we could never replace *Trygg*, but John-Andrew was keen to continue. Like me, he was enjoying the work so much that he did not want it to end.

After identifying suitable boats I slipped out of the Haa to walk to Elizabeth's to make some phone calls. I knew of a boat in Shetland that would suit us, although it was not advertised for sale. I had met the owner on Barra and so gave him a ring. I found out he was down south, returning to Shetland the following week, but I spoke to him and explained our loss.

'Just take the boat and use it,' Tommy replied, in answer to my request about purchasing it.

'But we might lose her like *Trygg*,' I said.

'Don't worry about it, you'll not lose another.'

'I'll come across to Shetland when you're back,' I replied, putting the receiver down with hope reborn. I was delighted, but I knew we had to buy the boat not borrow it, as we would need to modify her to suit our work and we *might* lose her.

'Good news?' asked Elizabeth, as she saw my expression.

'I hope so,' I replied, as I thought how I would broach the subject with Simon.

'Harry and I were sayin' how much we'd miss you all if you left.'

'No fear of that.' I smiled at Elizabeth.

It was only a day since we had lost *Trygg* and I had discretely tracked down a boat that would be suitable, and another two in the *Fishing News* that were too big, but they might help to persuade Simon to go for the smaller one. Moya gave me great support but was careful not to get too involved. She walked a tightrope: she had become an integral part of the business and yet we both had to consider Simon. With the loss of *Trygg* we now had to adapt and change, tearing up old plans and dreams. She coped with ease, seeing it like I did as an opportunity, although I know I behaved badly, as I thought of little else and sleeping was difficult, tossing and turning all night, my mind unable to relax. I had become unknowingly short and sharp when answering questions.

With our experience gained on the *Oceanic*, I was sure we could improve on *Trygg*. We had sold loads of brass and had more than 80 tons stored on Foula, let alone what remained on the wreck. I was now attempting the purchase of a new boat against Simon's wishes, but I was convinced he would come round. During the rest of the week everyone became accustomed to the loss. The initial impact was fading and we worked hard tidying up the bits of wood, keeping anything of use we found on the rocks, and diving to recover the salvage gear for reuse, all the while licking our wounds and discussing the future. I began to sense that Simon was gradually accepting the purchase of a new boat – *as long as it wasn't too big*. The rubber inflatable was carried into the old shop, now disused. Using the paraffin heater, we were able to dry it out and repair the leaks where it had been scuffed and punctured. Simon started to sort out the insurance with Harry, the only person on the island who could be called an 'official', as sub-postmaster. He asked him to send a letter to our insurers explaining the circumstances of the loss and noting that *Trygg* was beyond repair.

Instead of the weather improving, it became worse, with more gales and driving snow. Any work outside was miserable, forcing us to sit inside repairing diving suits and generally being bored and restless, not knowing where our future lay. At night when the rain stopped the Northern Lights were clearly visible, but instead of gently dancing they looked wild, almost frightening in their movements. It was as if they were taking over the sky.

Nine days after the loss of *Trygg* we launched the inflatable and checked the outboard before leaving for Scalloway.

We went to see Andy Beattie, who was dealing with our insurance. Simon had all the paperwork prepared, making it easy to complete the claim. Leaving Andy's office, we went to look at fishing boats. The first one, the *Silver Chord*, was seventy feet long, slightly on the large side for Foula, but in good condition. She was laid up and was therefore immediately available. While we went over her, Simon said little until we had finished.

'What about the smaller one you mentioned?' he asked.

'She's in Hay's Dock,' I replied. 'Just a ten-minute walk.'

The *Valorous*, at sixty-two feet, was twelve feet longer than *Trygg*. Originally a wooden fishing boat, she had been purchased by the 'gut factory', a fish-meal factory, which had fitted her with a watertight steel hold and a substantial derrick with a grab to collect all the fish guts from the fishing fleet. We were told that the second choice of name by the factory had been the *Odorous*! When the factory no longer needed her, she was sold to Tom Clark, who owned a commercial diving business. At present he had no work for her, and she had been laid up. It was him I had phoned from Elizabeth's.

'What do you think, Sy?' I asked.

'She's a better size than the other boat,' he replied. This reluctant approval was all I needed and we set off to see Tommy at his home in Lerwick.

'Will you sell *Valorous?*' I asked.

'Just take her away and return her when you've finished,' he said.

'We have to pay you,' I replied, 'in case we wreck her like the last one.'

'I may need her for another contract,' said Tommy, who was undertaking civil engineering work on harbours. 'I'm not sure I want to sell her.'

'We'll pay you, and if you want her back you can have her for the same price,' I suggested, before adding, 'Or if you want to make something on it, we can sell her back at a lower price?'

'OK,' he answered, and we agreed the reduced price.

Simon had joined in the bartering and was happy with the result. We had ourselves a boat. There was no written contract, just a handshake and a glass of beer. This was Shetland.

I was delighted, hardly able to contain myself. She was bigger and would be a major improvement on *Trygg*. As a bonus she had a Department of Transport Load Line Certificate, along with certified safety equipment, all the correct certification that *Trygg* had lacked. We had a few minor alterations to carry out on the lifting gear, but a day's work would be sufficient. Luck had been on our side. It had only taken nine days to buy an improved replacement for *Trygg* and no diving weather was lost. We left *Valorous* safely tied up in Lerwick until the weather improved.

I became incredibly buoyant after our meeting. Everything was going well. I had no reason to suspect anything would go wrong. We had been playing safe by leaving *Valorous* until the weather was settled. The insurance might consider losing one boat an accident, but to lose another would be sheer negligence.

Valorous's first day on Foula was spent in preparation for the wreck. She was emptied of anything not required, including damp packets of food and three large bundles of dog-eared magazines that had made up the boat's library. Once the inside of the boat was tidy, Moya started painting the decks and rails to make her look as though she matched the certification she carried. John-Andrew continued to dive and work on the boat with us.

It took just one tide working on the *Oceanic* to confirm that *Trygg*'s loss had been to our advantage. *Valorous* could carry more cargo, as well as take bigger lifts over the side. With this increased salvage capability, every time we lifted scrap from the wreck, much more than we had lifted with *Trygg*, we would return to the mooring in the voe, where we would reorganise the recoveries in the safety of calm water. They were either carefully placed in the hold or the larger bits on deck dropped in the voe beneath us to make sure *Valorous* was stable and had the space to lift more on the next tide. On one of these occasions I was in the hold positioning the lumps of scrap as they were lowered down. We were nearly finished – only a large piece of brass casing remained to be positioned. It was awkward to sling and its buckled half-cylinder shape made it difficult to stow. Several lifts and lowers were required in an attempt to fit it in like a jigsaw part. I almost had it in place – it required one more lift to let me turn it a fraction – but when the wire was hoisted it snagged on the main steel hatch beam. I looked with horror as the eight hundredweight beam jumped out of its retaining socket and started sliding down the wire towards me. The beam was twenty feet long; there was nowhere I could jump to avoid it. The winch was stopped, but it was too late: the beam continued to slide down the wire. I put my arms out in a futile attempt to stop it, but it pushed me back, pinning me against the scrap by my waist.

I knew it was serious. I forced myself to use my free hand to take the hook off the sling before twisting it round the beam to lift it off me. It was a relief, as the beam came off, but I had difficulty moving. Simon and John-Andrew helped lift me out of the hold and lowered me into the inflatable. As we motored to the pier I could feel my mind wandering and the mists coming in. I fought hard to keep conscious, doing my best to assist as they took me up the steps onto the pier, where I collapsed beneath the crane and then the pain started to hit me.

The next thing I remember is hearing Moya. 'Alec, Alec, how are you? Oh, Alec,' she said in a breaking voice.

I could feel her take my hand but was unable to see her. The sound of her voice encouraged me, so I tried hard to speak, but all I could do was listen. I must have looked unconscious, as no one seemed aware that I could hear. It was as if I were in a dream, willing myself to wake.

I recognised the sound of the nurse's van stopping beside me. 'How is he?' Maggie asked. Moya said nothing. I could picture the scene now in my thoughts. Maggie's voice gave her position away, as she bent over me, then I could hear other people arriving. *Word must be getting round the island*, I thought.

Moya spoke, reassuring me, and I knew I had everything to live for.

Maggie's voice ordered the others to gently get me onto the stretcher and into her van. I felt hands gripping me. The pain was almost unbearable. Then I could feel myself being carefully lifted into the van.

'The ambulance plane should be here soon. I'll take him to the airstrip,' I heard Maggie say.

The car started. I could feel people either side of me. Moya was talking to me, as she held my hand. Unconsciousness was passing over me like the clouds passing over the sun. I was struggling to hang on. I felt the van start to move. As it climbed the steep track from the pier, I could hear people running behind.

'Hold onto the stretcher,' someone shouted. I remembered that when there was a coffin in the van the end stuck out the back, so the doors had to remain open. I could smell the exhaust. As the van hit the potholes in the track, the pain travelled in spasms throughout my body. I felt the road level off. *We must be passing Ham*, I thought. *Downhill now, Harry's post office will be on our right.* Two bumps as we went over the small bridge below the post office. I sensed the van turning. *We must have passed the kirk that I was married in.* At last, the sound of an aircraft when the van had stopped. The noise changed, meaning it must have landed and was turning. All became silent when the engines stopped. The van moved forward and then reversed. I could hear the pilot – it was Ian Rae, who had been at my wedding.

'We're lucky the weather was so good for landing,' he said. 'We took the seats out in Tingwall to make space for the stretcher.'

I tried to move to ease the pain but gave up, exhausted.

As they manoeuvred the stretcher into the plane I heard the pilot complain about his back. The door shut, and Moya's voice and touch reassured me.

The aircraft shook as the first engine started, then the second, before it moved slowly along the rough strip. The engines revved and I sensed the plane turn and start its take-off run, the gravel rattling off the fuselage, thrown there by the thrust from the propellers. As the speed increased and the bumps grew worse, up into the air we went. No movement; it was bliss. I listened to the steady drone of the engines as he throttled back to cruising speed. I was on my way. I was off the island.

Not long before we landed, I knew I was going to make it. I was feeling better. By the time we landed at Tingwall airfield, I could feel myself starting to control my movements. My speech had returned and I was able to communicate. At the hospital they tried to get a drip in my arm but it was covered in the black stain that came off the scrap. Finding nothing strong enough to remove it they resorted to using an abrasive floor polish.

The worst damage was the crushing of my kidneys. I had to sign the consent form for them to be removed before I was taken into the operating theatre for further investigation. Fortunately the doctors were able to determine only internal damage, with the kidneys outwardly intact. It was a question of time to see if their impaired function would improve.

In order to be close enough to visit, Moya stayed with Bessie, who was from a Foula family and lived in Scalloway. Initially I was just pleased to be alive but after the first ten days when the pain had subsided I became an unsettled patient, enduring sleepless nights and long days, only broken by Moya's visits. I longed to get back home to Foula. I pleaded with them to let me go.

At last the day came when I was to be released. I felt elated at leaving the hospital. Simon picked us up at the airstrip on Foula and before going to the Haa, much against Moya's advice, I went down to see the engine of John-Andrew's new lobster boat, which had been giving him trouble. I felt so well being back – I felt invincible, ready to start work immediately, but Moya was right. I quickly began to feel light-headed and had to leave the boat and go up to the Haa. I would need time to recover.

12

Risks and Success

My health improved, and although I was unfit to dive, Moya and I continued our walks along the *banks* (coast), which we had always enjoyed, lying on the grass when I was tired, picking up drift-wood, floats or anything of interest that was in accessible places. We couldn't get enough of being together and the walks gradually became longer until I was physically able to revisit any part of the island. At the south-west we could climb down the Sneck of the Smaalie, a rock fault formed by the parting of old red sandstone cliffs. The Smaalie is said to be a troll-like creature and the Sneck is a sheer-sided crack about 200 feet deep, seven feet wide and 300 feet long. It had some large boulders jammed in it near the top to form a bridge that was both narrow and concave, making it a risky crossing. Over the centuries, young Foula men had crossed over it as a dare, or a right of passage. The only access into the Sneck was from the landward side down a grassy slope before it turned into an awkward, wet and slippery climb over collapsed stones smeared with green algae that covered much of the stone walls at either side. Drips of water and running streams from above gave life to ferns and damp-seeking lush vegetation which had managed to get a foothold on the vertical sides; it reminded me of a tropical house in a botanical garden. Once submerged beneath the sunlight in its dark, cold and eerie depth, the water dripping down the sides splashed onto our oilskins. Small movements, like an unsuspecting seabird flying off, would make my heart miss a beat, its sound reverberating within the cleft. At the very bottom it was wet and spongy underfoot, and we stepped warily over piles of bones and rotting carcases of dead sheep and birds that had fallen in. It was easy to imagine that the Smaalie was still living there, catching and feeding on anything that came within its grasp. I felt relief as the light increased when we approached the west end, where a rocky slope descended into the sea. On our right was a cliff face that led to Wester Hoevdi, rising 800 feet from the water. The only safe way back was to return the way we had come through the Sneck.

Once out of the Sneck we would take off our oilskins. Sometimes it was warm and windy, but we'd find a sheltered spot far enough away to be out of mind and reach of the Smaalie. Lying on the soft grass, mown flat by the sheep, we would become entranced by the peace and solitude that was enhanced by the sounds of the birds. Watching the puffins returning with sand eels in their beaks, the fulmars gliding in the updrafts, it was difficult to think back to the loss of *Trygg* or my accident and believe they had happened. My life was back in perspective. This was paradise again.

One day, as the sun disappeared behind the clouds and I could feel the temperature drop, I felt reluctant to move, not wanting to break the moment. It was forced upon us when the clouds finally looked as though they were there to stay. We walked slowly back along the side of the Daal, the glacial valley shaped like the bottom of a bath, with hills rising steeply from either side, the wet mossy patches concealing damp holes that could take us down to our knees. We carefully followed a sheep track that took us through the hidden nests of Arctic Skuas and Great Skuas. I grasped Moya's hand, raising my other hand above my head, breaking into a run to get clear of the birds as they dived on us, intending to hit us in their successful attempt to drive us away.

Slackening our pace after we passed them, we sat on a rocky outcrop as the sun slowly reappeared from behind the clouds. We could see the Shaalds, the water tumbling over itself in the swell. Behind the Shaalds in the far distance was the outline of Fitful Head on the Shetland mainland. I thought of the rest of the world as a shadow in the distance, almost an irrelevance. The passing of laws and actions of governments had little consequence on this island. We chatted about living on Foula; the *Oceanic* would not last forever and the loss of *Trygg* had concentrated our minds on the time ahead.

'I wonder why the birdwatcher's hostel was never completed near Ham,' said Moya. We had seen the weathered remains of foundations on Ham croft and been told of its purpose.

'It would benefit the island,' I said.

'On Fair Isle, some of the birdwatchers bring their partners to enjoy the island but the accommodation and meals are good,' offered Moya, pointing to a small lump on the horizon that could be Fair Isle, or perhaps it was a fishing boat. There was silence as I thought of the benefits of a hostel and yet I knew that it was not something I wanted to do. Neither did Moya. But we felt someone ought to. The mailboat would get more passengers, the shop could be renovated and it could be a showcase for small items made on the island, like Jock's letter openers and lamps made in his forge. Eric was a good artist. He might sell some of his drawings and paintings, along with cured sheepskins. It could be the start of a change: the increased number of people we had seen passing through the island since we had been here made it more likely someone might wish to marry an islander or settle on the island. The island could not be sustainable with so few people; it had to have incomers.

'Some of the islanders would object if hordes of people came. It would change the nature of it,' answered Moya, watching the sun disappear behind the clouds again.

'Yes,' I mumbled, 'but it's the price they might have to pay for a lasting community. It might also help solve the aircraft problem and make it sustainable.'

'Have they made a final decision yet or do the flights get questioned every year?'

I knew that the Shetland Islands Council had wanted to stop the ambulance flights, leaving the Aith lifeboat to undertake the service. Every household on the island had written a letter to say how essential it was for the older people. We had also sent an identical letter but added that, as divers on the island, it was reassuring to know the service was available. The islanders received polite replies; we received a very different one:

Dear Mr Crawford and Mr Martin,

I thank you for your letter of 23 May, and have to say that I appreciate your concern.

However, I am sure that you will agree that the Shetland ratepayers should not be expected to pay between forty and fifty thousand per annum to provide protection for persons choosing to pursue a hazardous pursuit in their area.

Fortunately the flights continued, or I might not be alive.

'Come on, then,' I said, as I eased Moya off the ground before setting off at a slow run along the sheep track. The Daal opened up to the Hametoon, where we walked through the ruins of three crofts: Stoel, Guthren and Crugalie. Sheep were using them as shelters, the roofs long gone and the walls crumbling. We poked among the fallen stones, looking for hints of the past, before passing the front of Jock's house, North Biggins, and the Isbister's South Biggins, which lay next to it, making our way in the general direction of the ruined church. The slate roof had fallen in and there were fulmars nesting in the exposed stonework. They cackled a warning to us to keep away or they would vomit their toxic oil. Retreating from the birds we closed the gate and saw the Isbister family working in their kale yard at the side of their house. They stopped their work and waved us in for a cup of tea.

There were no distinct boundaries between work and pleasure; they ran seamlessly together to mould their way of life. The weather, peats and *Oceanic* were all discussed, along with polite curiosity about *Valorous* and gentle questions on the next trip south. Aggie-Jean, Bobby and Eric were a happy family, but the days of Bobby's Foula and his life aboard sailing ships was long past. Now in his late seventies, they cut their peats for fuel, used Tilley lamps for light and drew water from a well, relying on the old-age pension, sheep subsidy and growing their own vegetables to survive. While Moya discussed the lambing with Aggie-Jean, I looked out of the window at the land that had originally been cultivated and was now grazed by sheep. Rushes and irises had taken over the wetter parts – forty years ago that land would have grown a coarse grain to be ground and used as flour. A water-powered mill lay ruined at the lower end of the Toon, and six derelict crofts – Norther House, Shoadals, Bankwell, Grisigarth, Quinister and Newhouse – could be seen from the house. Out of my sight were the ruins of Breckans, which had been the last blackhouse in Shetland, where the fire was in the middle of the room, a hole in the roof to let the smoke out. The occupant had died less than 15 years previously.

I half listened to Bobby, who with his limitless memory told us of the various names that had gone down through the island families. His ancestors were adventurous, working at sea and travelling around the world. I had realised when I met Moya that my dreams of building up an old croft were just dreams. Moya was ambitious, like me, wanting to do something special with her life, and to that end we were planning to stay in Fife when the work on the *Oceanic* was finished.

But I wondered about some of the present population: did they want to be on Foula, or if they had the same croft or an opportunity on the mainland, as I did, would they prefer to be there? The island had taken such a battering in the past few years, with marriages breaking and people leaving, it made it difficult for it to recover. Those who remained were left to muddle on with traditional croft work that was no longer sustainable without pensions or government subsidies. Some of the four schoolchildren might stay; others would leave when they started their working lives. As they were all related, they would be looking for partners from outside the island.

I sensed Bobby saw it as a fait accompli, as if there was nothing he could do: he would live his life out on the island and that was it. I could never see Foula uninhabited, but I wondered to what level the population had to fall before a schoolteacher, nurse, mailboat – all the facilities provided by the outside – were reviewed. Simon often remarked I was 'hungry for work'. Few seemed hungry in terms of actually doing something to make the island more viable, though perhaps John-Andrew's new boat might help. The rebirth of the island, if it were to happen, would require more than the efforts of one family, but there seemed a natural resistance to change. It was similar to my father's attitude when he would tell me, 'Your mother and I did it like that, so there's no reason why you shouldn't.'

Moya nudged me to bring me back from my reverie. I turned my head from looking out the window to join in the conversation. Reluctant to leave the warmth and happiness of the Isbisters' house, we stayed a while longer before we headed towards the Haa. We would both miss Foula when we left: the feeling of risk on the Shaalds and the happiness ashore would be hard to beat.

When we reached the Haa, Simon was lying on the settee with

his bobble hat on, listening to sport on the radio. Exactly as we had left him.

'Anything happened?' I asked.

'Not much,' replied Sy. 'The "Provider" was in and left a few fish for us.' There was a pause as he raised himself off the settee 'They're in the sink.'

Simon and John-Andrew had continued the work on the reef with the inflatable. *Valorous* lay moored in Scalloway until I was fit enough to operate her. Our first job on my recovery was to shift the two central propeller bosses, which we had lifted off the reef and taken into the shelter of the voe. At twelve and a half tons each they were too heavy to lift into the hold; they would have to be suspended beneath *Valorous* all the way to Scalloway, where they could be lifted out by a crane. These were to be the heaviest lifts we would make; they were also extremely valuable. When we were blasting a small chunk broke off and the analysis proved they were made of approximately 88 per cent copper, 10 per cent tin and 2 per cent zinc.

The trip to Scalloway with one hung beneath the boat was nail-biting. As Simon put it, it was all the company profits dangling on a thin wire. If the wire had broken, they would have fallen into water too deep for diving and the bosses would be lost. One of us regularly hung over the bow to check the wires didn't chafe, but otherwise there was little we could do. *Valorous*, being bigger than *Trygg*, did not go down so much by the head and with some scrap in the hold to ballast her she felt safe, managing the job with ease.

Regular trips were required to transport the smaller scrap south, where we loaded it onto lorries at Scrabster. It was our nearest Scottish mainland port at 120 miles, taking around 13 hours to get there.

The *Oceanic* was coming near the end of its profitability – scrap remained but it would soon become so scarce that it would be uneconomic for us to lift it. We had done well, set ourselves up in life: Simon was looking at buying a pub or a hotel in St Andrews, John-Andrew had bought a lobster boat to work off Foula and was now thinking of buying a small trawler, while Moya and I were going to take over a farm in Fife, but we both intended to continue with salvage and had been researching wrecks that could be viable.

While unloading at Scrabster we met Denny, a likeable man who was the skipper of a trawler called *Prolific*. Between whiskies in the Captain's Cabin, he required little encouragement to tell us about the area, regaling us with tales of the island of Stroma, on which some of his family had lived, including detailed descriptions of the shipwrecks that had occurred. The area is well known for the strength of its tides, with measured speeds up to 16 knots in the worst places, causing whirlpools and over-falls, where the water is pushed up by the fast movement to make it look as though it is going over a weir. In bad weather the force of the water and large waves can quickly change the heading of a ship, resulting in casualties on the rocky coasts of Stroma and nearby Swona.

Stroma, the most southerly of the two islands, lay 14 nautical miles from Scrabster in the centre of the Pentland Firth, two miles north of John o' Groats. Keen to look at the wreck sites, I set off at 10.30 a.m. in the inflatable, leaving John-Andrew on *Valorous* as Simon and Moya had travelled south with the scrap. I took a compass, flares and foul-weather gear, along with spare fuel and a set of tools. Unfortunately our spare outboard was on Foula. It took me over an hour before landing at the small harbour at the south end of Stroma. The island was low-lying and flat, about two miles long and one mile wide, but ringed by cliffs that were about 100 feet high on the west, with the ground sloping down to the east, where the cliffs looked like a high wall with a rocky foreshore. It was difficult to believe the island had supported 375 people; now there was no one, the last twelve people had left in 1962. I looked inside the complete but uninhabited houses. It had been prosperous at one time. The houses were well built with good quality stone and slate roofs, unlike most of the Foula houses. Sheep had wandered into the ruins for shelter, depositing the mud from their feet, allowing grass to thrive near the doors and broken windows.

Overall, it was a sad experience; I thought of Foula and hoped it would never come to this. The remains of human habitation made the island feel unwanted and unloved, no tilling of its soil, no children shouting or playing games, a feeling of emptiness. I didn't like it, I felt uncomfortable. Walking towards the north-west of the island, I was compensated by my fascination with a partially collapsed sea cave called The Gloup. It is a large hole surrounded by cliffs, filled with seawater and connected to the sea by a subterranean passage more than 100 yards long. Denny said it was originally used for smuggling, as it was only accessible from the sea at low tide, as the passage was not high enough to take a boat when the tide rose.

Making the two-and-a-half-mile crossing from Stroma to Swona, I passed through strong currents, over-falls and whirlpools before reaching the lea of the island of Swona. This island was just over a mile long and about half a mile wide. Looking at my chart, I tried to find a safe landing place; most of the cliffs were on the east side, but there was a beach marked and I headed towards it, as there was too much swell to land on the west. The water splashed over my boots as I waded ashore on a boulder beach. The ground looked poor, it was like hill land with thin soil and rocky outcrops. The island had only ever supported a maximum of thirty people, becoming depopulated in 1974. Denny had told me of the *Johanna Thorden*, a Finnish vessel that was travelling from New York to Gothenburg in Sweden on 12 January 1937 when hurricane-force gales drove her aground on the south-western part of the island with a cargo of copper. Making my way to the wreck site, I had a sudden fright, unexpectedly coming across someone.

Catching my breath, I said, 'Hello.' Then feeling I should explain myself, I added, 'I'm having a quick look round.'

'I'm over to check my stock,' the man replied.

'D' you live here?' I asked.

'No, I've a farm on South Ronaldsay. I've a few cattle and sheep on the island.'

We talked for a while before he pointed me in the direction that would take me to the wreck site. I hurried on, as I had already spent too much time on Stroma and now it would be the evening before I would get back to Scrabster. The wreck site was on the edge of the strong current and did not look a good place to dive; it would be possible, depending on the depth. Returning to the inflatable, I motored at a slow run round to have another look from seaward before heading back to Scrabster. Relaxing as the inflatable flew across the water, I began thinking of having a pint with John-Andrew.

Suddenly there was a loud grinding noise and the outboard stopped with a jerk. I pulled the starter cord several times but the engine had become hard to turn and made a scraping noise when it did turn. Unscrewing the brackets, I lifted the outboard off the stern and set it on the floor of the inflatable. Using the tool kit, I started to take the flywheel off. Half an hour passed before I exposed the internal damage and realised it was unfixable without some spare parts. I cursed myself for leaving the spare on Foula, but this trip had never been planned.

Drifting westwards on the current, the inflatable was sheltered from the southerly breeze and moving in the right direction for Scrabster. I tried paddling but the effect was minimal. There was a coastguard hut at Dunnet Head lighthouse; when I drifted past it, I hoped it was manned. Searching for other boats, I saw a large ship several miles ahead of me heading in my direction. With my orange foul weather jacket tied to the paddle I waved it. I was in the middle of the channel, so there was no doubt they would see me. Continuing to head straight for me, the distance between us closed until I could see people on the bridge wings and recognised the ship as a Russian factory trawler. I was sure they saw me and yet the ship continued on the same course straight towards me. As she closed in, the large overhanging bow obscured their wheelhouse. I lay on the floor holding onto the grab ropes hoping the inflatable would not turn over.

The impact of the bow wave pushed the inflatable aside and washed it down the side of the ship. Looking up there was a group of sailors hanging over the side looking down at me. One even waved. I pointed to the lighthouse with one hand and held the other to my ear as if holding a phone. The ship passed, leaving me in their turbulent white water with just a fraction of hope that they would notify someone.

I continued to drift past Dunnet Head, noticing the inflatable was now moving north as well as west. I would no longer pass near Scrabster but was heading for the open sea.

As the hours passed and it was starting to get dark I saw the ferry leave Scrabster for Stromness on Orkney. She passed a few miles ahead of me – she was the only ship I'd seen since the trawler and I set off three flares. I had been facing their stern, but I was sure someone on the ferry would see them. I kept watching Scrabster harbour, hoping to see a fishing boat or the lifeboat making its way out of the entrance. I saw nothing, and now the darkness was rolling in, with damp air from the south-east wind. The depth of water was far too great for normal anchoring, but I made a sea anchor using the plastic fish box we used for storing loose items. Attaching it to the small anchor, I dropped them both twenty feet into the water. The drag it caused kept her head into the wind but I estimated the inflatable was drifting at a rate of two to three knots. There was nothing I could do. I just had to be patient and sit and wait. My life was in someone else's hands.

As I drifted into the ocean, the swell started to give the inflatable boat a new and uncomfortable motion. The wet fog intensified the gloom of the approaching evening. I listened for the faint blare of a distant foghorn only to resign myself to the fact that no one had seen the flares.

I was on my own in the Atlantic.

The evening passed slowly as I kept alert, watching for navigation lights or engine sounds, without hearing or seeing anything. My spirits and hope began to diminish. The cold, starting at my fingers and toes, worked its way through my body, slowly enveloping my legs and chest, eventually causing me to shake uncontrollably.

By the time I was completely immersed in the damp foggy darkness I felt afraid. The sea had always been my friend, but now the heavy dark swell slipping under the inflatable seemed hostile as it raised and lowered me with the unstoppable strength of each passing wave.

Time crept by. I longed for dawn to appear, when there would be a better chance of being rescued. I thought of how to react if the boat was run down by a ship or driven towards the cliffs on the north of Scotland. Would I survive the cold of the water if I was to swim clear and take my chance of finding ledges to climb, knowing the heavy swell would try and sweep me off? Past experience had made me wary of diving near partly submerged rocks, as the barnacles and sharp edges soon destroyed a diving suit.

I began to feel hungry. I thought of drop scones. I had seen the ingredients in *Valorous's* galley when the stores had come aboard. Flour, eggs and sultanas, with some fresh milk that had been set aside next to the cooker; they tasted better with the fresh milk than the powder we often used. I imagined them lightly cooked with the centres soft, the butter melting over them – jam or marmalade spread on top. Everyone liked them; they were used in our house as a stopgap between meals when the islanders came in to wait for the mailboat. The girdle, with its half-moon-shaped handle, lay on the open peat fire beside the kettle. The mix was poured on it and the cooked scones were scooped straight onto our plates – or into our hands as we waited for Harry to appear at the door and shout, 'Boys, I see the boats coming!', then we'd all rush down to the pier, often carrying a half-eaten scone.

The judder of a wave snapped me out of my dream, but I could have almost smelt the scones – with lashings of marmalade. Yes, definitely marmalade.

Moya must have heard by now that I hadn't returned, I'd let her down, but I knew she would make sure all the fishing boats would be notified over the radio, the lifeboat would be launched. They'd not be able to use a helicopter in this weather until daylight.

My thoughts were disturbed again by the violent slop of a rogue wave pushing its broken crest over the side of the inflatable. Hanging onto one of the grips to stop myself from being thrown over the side, I bailed her out again before I returned to sitting on the floor, my legs braced against the opposite side to keep me steady and give the boat a low centre of gravity. At times there was a momentary pause in the noise of the waves as they found their rhythm, the tops no longer breaking. A low-hanging silence would then surround me, so quiet I was conscious of the sound of my breathing.

I blamed myself for what had happened. Yes, I had cut corners, but we'd never have succeeded if we hadn't taken risks, financial and physical. It had been a struggle, but eventually luck had come our way. Perhaps I had become too confident, not realising I was subject to the same fate as anyone else who makes a mistake? For years I retained that confidence of youthful immortality – *it won't happen to me* – but now it was wearing thin.

I soon became aware again of the chill taking over my body. Movement made little difference, as I would feel the warm air closest to my skin escaping through my woollen jersey and water-proof jacket. Curled up on the floor was my warmest position, but I knew I mustn't fall asleep. Hypothermia was a distinct possibility. I must sleep during the day when it would be warmer, not at night.

At last five o'clock came, the night was nearly past, only an hour before this damp eerie darkness would start to lift. God, I was cold. I regretted setting off all the flares and wished I'd taken a torch. I didn't even have matches.

I looked at the compass. I was drifting on a steady westerly course. I thought, *I must be about eight miles off the north of Scotland and twelve miles west of Orkney.* This was an estimate since I had not seen land for ten hours. With the freshening wind and increasing swell, I could be anywhere.

I thought of all of our dreams and plans. Surely, it wasn't going to end out here. How on earth did I get myself into this? Where had I gone wrong?

As I drifted off in the cold I could imagine Moya joining me in the front of *Valorous's* wheelhouse, using her elbows against the bulkheads to steady herself against the motion.

Shaking myself awake, I knew I mustn't fall asleep. I was too cold, it wasn't safe. I looked at my watch again. Only fifteen minutes had passed.

My ears pricked up. In the slowly approaching dawn I could hear something. In this mist-driven morning I knew from our own radar how the screen became 'cluttered': a heavy swell with breaking water would blur small objects, making them difficult to recognise, but the visibility was opening up to about half a mile. I shouted, knowing there was little chance of anyone hearing me above the noise of a boat's engine, but I heard the sound again. I stared in its direction when the inflatable was on the top of a wave. It was definitely an engine. Disregarding the cold, I took my orange jacket off, forcing the paddle in one arm and stood up. Trying hard to keep my balance each time, I waved it above my head when the inflatable rode on the peak of the swell. I was sure there was something there although puzzled because I couldn't see a mast. Perhaps it was just my eyes, but the noise was becoming consistent.

Exhausted by the waving and unable to keep my balance I sat on the pontoon for a few minutes, gripping the rope along the side to keep me steady, still concentrating my eyes in the direction of the sound. My heart beat harder: this was it, I was going to be rescued. On the next high wave I was sure there was a figure standing on the bow of a boat. The boat looked small; it might be the Thurso lifeboat. Standing again, waving the jacket, I saw the person waving back. Relief flooded over me. Thank God, thank God, I thought, as all the negative feelings vanished from my mind.

The lifeboat was next to me. It was John-Andrew standing at the bow; he looked as happy as me and helped me aboard. Going below, the warm air washed over my face, while my cold hands were hardly able to hold a mug of tea. I had been lucky, thanks to John-Andrew, who had insisted they continued going further out to sea. I was told he had stood at the bow the whole trip to look out for me. It was typical of his nature.

We could sense each other's pleasure at the outcome: there was little need to speak, but I realised my life had suddenly become less predictable. I wasn't so sure of anything.

When we landed at Scrabster, I was reassured to see Moya. It was as if I could not believe she would be there. Everyone was telling me to go to my bunk, but I desperately wanted to sail home to Foula. We left Scrabster at 10.45 a.m. the sun had burnt the fog away and the breeze had dropped, leaving smooth water on top of a heavy comfortable swell. As the mainland disappeared in the distance, a feeling of absolute relief swept over me. I had felt foolish at being adrift at sea, but now, at the wheel of the boat, my confidence was returning. In the back of the wheelhouse I could hear the sizzle of drop scone mix as it was poured into the frying pan. Simon and Moya were talking; John-Andrew was in his bunk. At last Moya appeared with a plate of food and two mugs of coffee. She was using her elbows to steady herself against the swell. I smiled.

It was after midnight before we secured *Valorous* to her mooring at Foula; John-Andrew left for home on his motorbike, Simon went to bed, while Moya and I sat in front of the blazing peat fire in the sitting room, not bothering to light the Tilley lamp. Sitting there together, unwinding after a very long two days, I realised how lucky I was and how valuable was this gift of life that I had nearly lost. Perhaps my life had to change.

The following morning Elizabeth banged on the open door and shouted in, 'Thy mother phoned last night, you'd been late back or I'd hae seen the boat lights and come an' fetched thee.'

'It was after midnight. I'll be up to phone her shortly,' I replied.

My mother was pleased to speak to me, but sounded concerned. I could feel her relief when I told her that we were *all* back on Foula.

'Your father was invited to a Forces party at RAF Leuchars two nights ago,' she said. I then heard my father in the background shouting a message for me: 'Tell him when I mentioned Scrabster they told me they planned to send a search and rescue helicopter out in the morning as some *idiot* was adrift in a rubber boat in the Pentland Firth.'

I emphasised that we were all safe.

I wasn't sure she needed to know anything else, but it did appear in the press the following day under the headline 'Pentland Firth Search – Man Found'.

The *Oceanic* was like a mineral mine that had been worked out: it was time for us to move on. The scrap we had recovered would be melted down and recycled; it might reappear in car radiators, water pipes or even bronze sculptures; bits of the *Oceanic* would go on for centuries playing a role in peoples' lives without them knowing. We had kept a few interesting pieces and the Shetland Museum and Ulster Museum each had a propeller blade to put on display.

We continued blasting with the inflatable when there was too much motion on the wreck site for *Valorous* to work, but gradually the conditions deteriorated to an almost-continuous heavy swell that made it too dangerous even if there was no wind. As the swell was likely to last for a few days we started the easy job in the voe of lifting bundles of condenser tubes into *Valorous's* hold. I carried out much of the diving – it was only in twenty feet of water but it was repetitive, continually coming up to the surface and going down again. We recovered most of the tubes but on the third day of this shallow work I began to feel a bit rough and was glad to go ashore for an early lunch. I had already cost us working time by being injured earlier in the season and was reluctant to hold us up any more, but my condition quickly deteriorated; I became dizzy, unable to stand and finally physically sick. Helped up to my bed as I had no balance, I lay down but could no longer tolerate the light. Moya became concerned and the nurse was called. It was possible I'd had a stroke. The dreadful procedure of getting to the ambulance plane started again, although this time I lay in the back of the nurse's van with a bowl in my hands and my eyes tightly closed. As I was lifted into the aircraft I could hear the pilot say, 'Not him again! I hurt my back lifting him in last time!'

By the time we made the short flight to Tingwall and slid into the waiting ambulance, I was starting to feel slightly better. The doctors were unable to diagnose a reason for the problem, and as I quickly improved they let me leave after a few days, but I knew my youthful immortality was slipping away. Moya had been more concerned over this incident than the previous one, as there was no obvious reason for it. I think we both knew that I had been lucky to survive that year's work and should stop diving for the present. It was suggested I should see a hospital when we went south. It was not until a few years later, when I was on a medical course for divers, that the diving specialist suggested I had had a bend. They were discovering that heavily repetitive shallow diving, where the diver is continually breaking the surface causes a 'topping up' effect of the air being breathed, resulting in the diver suffering a cerebral bend, this causes the symptoms I had suffered. Cerebral bends are often fatal; I had been lucky.

Back at the Haa I was exhausted, but it was near the end of the season; we quickly loaded *Valorous* and sailed to Scalloway to fill her with fuel for the journey. John-Andrew had his lobster boat to make a living and Simon was still looking at the adverts in hostelry magazines for a pub in St Andrews. Moya and I knew that we needed more than a small farm to fulfil our lives and a chance conversation with Andy Beattie of Hay & Company was to influence the direction in which we would go. Andy had invited us to a meal with his wife Edith in the Lerwick Hotel when we were tied up in Scalloway; he always took an interest in our work and often asked us out if we were in Shetland. He had immaculate manners and was always smartly dressed, making us search for our cleanest clothes and be on our best behaviour. During a conversation about the BP fuel agency that Hay's operated, Andy made a comment that intrigued me.

'We used to have an interesting ship taking bunkers [fuel] at Lerwick. It was involved in salvage work off the Faroe Islands.'

'What wreck were they working?'

'I don't know,' replied Andy, 'but the wreck was carrying a cargo of Cornish tin. The salvage ship was called the *Droxford*.'

The wreck was the *Hollington*, which lay about 14 miles south of the Faroe Islands, about 180 nautical miles north and west of Foula. During the First World War she was on a trip from Liverpool to the White Sea, which lies east of Murmansk in Russia, carrying a cargo that included thousands of army uniforms, a large quantity of live artillery shells, several million rounds of small arms ammunition and 718 tons of Cornish tin. Torpedoed in June 1917 by the submarine *U-95* she was lost with thirty men, including the captain. During 1970 the *Droxford* found the *Hollington*, then, laying a mooring system in a depth of 894 feet, they recovered an initial twenty tons of tin. By 1973 they had recovered 646 tons. The *Droxford* then sailed to Newfoundland to recover gold from the *Empire Manor*.

I had read in books about Risdon Beazley, a leader in cargo recovery from wrecks, and envied the type of work they carried out. They were shrouded in secrecy – so tight that Simon and I used to joke that the confidentiality agreement with the crew involved getting your legs broken if you breached it!

I started dreaming of a more radical system, underwater cameras, hydraulic grabs and all the winches remotely operated by one person from the wheelhouse. It would be safer, more efficient and the salvage system could be operated by one person, not requiring any divers and the depth we could work would be unlimited. I discussed it with Moya; we had both agreed to buy out Simon's share in Crawford and Martin, as he already had his eyes on a pub. Our plan for the future involved farming the small farm in Fife but also to use it as the base for our salvage business.

'You won't need to dive any more if we move into deep water,' Moya said, looking relieved at the thought.

I hesitated. 'Um, well . . . Maybe one more diving wreck.'

I was thinking of a final shallow wreck where there was more than a thousand tons of armour plate, a valuable nickel chrome steel that had a very low radiation level. If we could develop a low-cost method of lifting it, like on the *Oceanic,* it could be quick and easy money. It lay on a reef known as the Bell Rock, 12 miles off the Fife coast.

Epilogue

A Family in Deep Water

The *Oceanic* provided the foundation for our future. Simon married and purchased a pub and wine bar in St Andrews; John-Andrew had success with his lobster boat and bought a small trawler, allowing him to go further afield with his fishing. Foula continues with a similar number of islanders and four pupils at the school. Moya and I moved into our small farm, using it as a base from which we could operate our salvage business. *Valorous* was replaced within a few years by *Redeemer*, a ship that had accidentally sunk off Lyness in the Orkney islands. Her hull sustained little damage, making her ideal for refitting in Dundee.

After trials in the UK, she worked abroad in the Atlantic and Mediterranean, equalling and then tripling the world record for deep-water cargo recover while using a steel wire rope. Finding the steel wires too heavy for use in deep water, we were required to become pioneers in the use of fibre ropes. Most of them are weightless in water, have a similar strength to wire rope and can be used at any depth without a weight penalty. Their use enabled us to operate in depths of 9,800 feet (3,000 metres) on wrecks and gave us the potential to go deeper. Work was also carried out on shallow-water wrecks around the world.

Moya ran the business, and when *Redeemer* became too small she charted larger ships, but we continued to develop and build much of the equipment that was required to work at these extreme depths, always with the intention of keeping down operating costs. Whenever Moya had spare time she would leave the office and work on the ship, operating the winches or helping wherever she was needed, knowing that if the business was to progress we both had to understand any problems that occurred during the actual operations.

After being contacted by oil-related companies interested in using our technology, which now included specialist winching equipment with enormous winches that could lift more than 100 tons in deep water, and which could also be combined with systems supplying power and fibre-optic controls, we moved into that field.

The salvage work continues. We have developed a deep-water oil removal system to combat pollution from wrecks, and deck plate removal tools to crush the steel plates on wrecks, allowing access to inaccessible areas – explosives are ineffective in deep water.

Life in the marine industry has never been easy, resulting in financial highs and deep troughs, but it has always been interesting and challenging.

Our four children, three sons and one daughter, are all involved with the marine industry.